Understanding National Identity

We live in a world in which being a 'citizen' of a state and being a 'national' are by no means the same. Amidst much scholarly debate about 'nations' and 'nationalism', comparatively little has been written explicitly on 'national identity' and a great deal less is solidly evidence based. This book focuses on national identity in England and Scotland. Using data collected over twenty years it asks:

- Does national identity really matter to people?
- How does 'national identity' differ from 'nationality' and having a passport?
- Are there particular people and places which have ambiguous or contested national identities?
- What happens if someone makes a claim to a national identity? On what basis do others accept or reject the claim?
- Does national identity have much internal substance, or is it simply about defending group boundaries?
- How does national identity relate to politics and constitutional change?

DAVID MCCRONE is Emeritus Professor of Sociology in the Institute of Governance at the University of Edinburgh and a Fellow of both the Royal Society of Edinburgh and the British Academy. He has published *Understanding Scotland: The Sociology of a Nation* (1992 and 2001); *The Sociology of Nationalism: Tomorrow's Ancestors* (1998); *National Days: Constructing and Mobilising National Identity* (2009, with Gayle McPherson); and most recently *The Crisis of Social Democracy in Europe* (2013, edited with Michael Keating).

FRANK BECHHOFER is Emeritus Professor of Social Research at the University of Edinburgh, Honorary Professorial Research Fellow in the Institute of Governance and a Fellow of the Royal Society of Edinburgh. He is a co-author of *The Affluent Worker* (Cambridge University Press, 1968 and 1969); *Principles of Research Design in the Social Sciences* (2000, with Lindsay Paterson); and also *The Petite Bourgeoisie: Comparative Studies of the Uneasy Stratum* (1981, edited with Brian Elliott).

Together they have published many papers and books, including *Living in Scotland: Social and Economic Change since 1980* (2004, with Lindsay Paterson) and *National Identity, Nationalism and Constitutional Change* (2009). They have a national and international reputation for their work on national identity.

Understanding National Identity

DAVID MCCRONE AND FRANK BECHHOFER

CAMBRIDGE
UNIVERSITY PRESS

University Printing House, Cambridge CB2 8BS, United Kingdom

Cambridge University Press is part of the University of Cambridge.

It furthers the University's mission by disseminating knowledge in the pursuit of education, learning and research at the highest international levels of excellence.

www.cambridge.org
Information on this title: www.cambridge.org/9781107496194

© David McCrone and Frank Bechhofer 2015

This publication is in copyright. Subject to statutory exception and to the provisions of relevant collective licensing agreements, no reproduction of any part may take place without the written permission of Cambridge University Press.

First published 2015

A catalogue record for this publication is available from the British Library

Library of Congress Cataloguing in Publication data
McCrone, David.
Understanding national identity / David McCrone and Frank Bechhofer.
    pages   cm
Includes bibliographical references and index.
ISBN 978-1-107-10038-1 (hardback) – ISBN 978-1-107-49619-4 (paperback)
1. National characteristics.   2. Nationalism.   3. Identity (Psychology)–Social aspects.   4. Group identity.   I. Bechhofer, Frank.   II. Title.
JC311.M393 2015
305.8–dc23
2014043721

ISBN 978-1-107-49619-4 Paperback

Cambridge University Press has no responsibility for the persistence or accuracy of URLs for external or third-party internet websites referred to in this publication, and does not guarantee that any content on such websites is, or will remain, accurate or appropriate.

Contents

Tables

Preface

This book is the product of a thoroughly collegiate form of working in which the data, the analysis and successive drafts, as well as the many papers on which it is broadly based, have been discussed and amended by both authors throughout, and they are equally responsible for it. David McCrone took responsibility for writing the first draft. We are both deeply committed to collegiate research but it is inevitable that one can only do this for twenty years or so by also being firm friends. Given our differing personalities and intellectual interests and strengths, our colleagues may well regard our enduring intellectual partnership as something of a miracle but we have greatly enjoyed the experience and that is what has enabled us to produce this body of work. The usual stricture applies. We and we alone are responsible for the research we have done together and what we have written; the faults are ours alone.

We have a lot of people and organisations to thank. By far the major funder was The Leverhulme Trust without which most of the research would simply not have happened. Successive directors Barry Supple, Richard Brook and Gordon Marshall were always supportive and understanding, and above all, wonderfully non-bureaucratic. We commend the Trust as an ideal funding body with which to work. We are also indebted to the Economic and Social Research Council which funded the arts and landed elites study and helped with funding Scottish and British Social Attitudes Surveys at crucial moments. The National Centre for Social Research (NatCen) which carries out the British Social Attitudes Surveys, and the Scottish Centre for Social Research (ScotCen) which is responsible for the Scottish Social Attitudes Surveys, have been especially helpful and encouraging. We particularly wish to thank Simon Anderson, Rachel Ormston, John Curtice and Susan Reid at ScotCen for their robust and incisive help with the social surveys, and their willingness to let us try out innovative ideas even if they were initially sceptical.

There are many colleagues to thank. Pride of place goes to our main research officer, Richard Kiely, who worked with us on the arts and landed elites project, the Berwick study, and the Leverhulme programme in the course of which he carried out virtually all of the intensive interviews on 'nationals and migrants' across Scotland, and, being unable to drive at the time, learned and suffered without complaint the intricacies of public transport in so doing. We are grateful too to Gary West and Robert Stewart who worked with us as research officers in the early days of the research on the elites, and Sue Renton and Margaret McPherson, our secretaries on the elites and Leverhulme projects respectively.

We are indebted to our colleagues and friends at Edinburgh University who not only played a vital part in the Leverhulme projects, but have been generous with their time and advice; notably, Lindsay Paterson, Jonathan Hearn, Michael Rosie and Ross Bond. Lindsay Paterson deserves our especial thanks not only for that general intellectual support but for his uncomplaining and lucid advice when our joint statistical expertise was not up to some task. Outside of Edinburgh, we have benefited from working with social psychologists Steve Reicher at St Andrews, Nick Hopkins at Dundee, Susan Condor and Jackie Abell, then at Lancaster. The great pleasure of academic life is that it is a collaborative exercise in which sharing ideas, findings, and having disagreements, is of the essence. We are grateful, too, to the anonymous referees and reviewers who have provided helpful suggestions and comments, and to John Haslam at Cambridge University Press for his patience and advice. Even when we disagreed with them, it made us think harder about how to present our ideas more clearly.

Finally, the biggest debt we owe is to our partners, Jean Bechhofer and Jan Webb. This book would never have been written without their support. They have not only had to put up with our long-standing interest, an interest bordering on obsession with national identity, but in the course of the last year they have tolerated the excessive amount of time which writing the book has absorbed in our joint lives especially as we have been supposedly largely retired.

Introduction

It seems reasonable to assume that anyone who picks up this book does so because they are interested in understanding national identity. After all, that is the title of the book, and our purpose in writing it is to explain our approach to national identity and present the empirical evidence which has led us to understand it in the way we do. We must make it clear right from the outset that the book is *not* centrally, or even largely, concerned with constitutional change. We are well aware that to some people it seems self-evident that the two must be inextricably entwined. The way political events in the United Kingdom have been reported since devolution in 1999 and the tendency in the media to assume a strong association between national identity and matters of governance has served to encourage that belief. The book will appear relatively soon after the referendum on Scottish independence held in September 2014, which is why we wish right at the outset to say that it is not about the referendum, and after this Introduction that momentous event will largely disappear from view in these pages. There are two reasons for this. Our intellectual focus over twenty years has been on various aspects of national identity. We do, of course, have our views, both personal and intellectual, about constitutional change, but that is not what our empirical research has been about, and this book is concerned with that research. Second, the connection between national identity and people's political and constitutional preferences is considerably looser than one might expect. We do not want to mislead the reader, so it is necessary to emphasise that the title of the book accurately describes its contents, which is at it should be.

How did we first become interested in national identity? Truth to tell, it is impossible to say. One of us has lived in Scotland for all of his life and the other for most of it. If you do that, national identity is pervasive and yet implicit. We are fond of a comment by the novelist Willie McIlvanney that 'having a national identity is a bit like having an old insurance policy. You know you've got one somewhere

1

but often you're not entirely sure where it is. And if you're honest, you would have to admit you're pretty vague about what the small print means' (*Glasgow Herald*, 6 March 1999). We began using that comment because it captures the ubiquity of national identity, as well as its imprecision and flexibility; the fact that it is part of the culture, up for grabs, and yet negotiable. This is not, however, simply a book about Scotland, but about national identity in Scotland and England, because you cannot understand the one without the other, and in any case, all forms of social identity involve an 'other', whether explicitly or implicitly.

We are sociologists, and interpret the social world in that way even if our observations and analysis of that world are influenced by other social sciences. We also take a specific sociological view of national identity. As we see it there is such a thing as 'society', but, despite the constraints of social structure, people are remarkably good at making sense of who they are for themselves, and acting on that basis. That, certainly, is one of the most important things we have learned from carrying out empirical research on national identity over twenty years. We have listed in the Appendix the key publications from a series of research projects and programmes so that the reader can, if they so wish, follow how our ideas and techniques developed. These publications, mostly in academic journals, set out the methodology and analytical techniques we have used and refined over the years in greater detail, especially statistical detail, than is appropriate in book form. Here, we have focused on giving the broader picture, on setting our results in a wider context. Looking at national identities in Scotland and England has implications going well beyond these islands, and because we live in a situation where 'national identity' and 'citizenship' are not commensurate with each other, this comparison provides an opportunity for a revealing and powerful research design.

We began exploring national identity in the early 1990s with a study of landed and arts elites in Scotland. We focused especially on 'incomers' holding positions which gave them some control over important aspects of the society into which they had come. They were confronted with claims that by virtue of being incomers they were poorly placed to understand that society. If McIlvanney was correct in saying that most people, notably 'natives' born and living in the country, are pretty vague about the small print of what national identity means, then 'incomers' often have to face claims that they are not 'one of us'.

We found that the study of such 'paradoxical people' gave us valuable insights into processes of national identification, and so, we argued, would the study of 'debatable lands', places in which national identities were ambiguous or multiplicitous. We followed up our 'elites' study with one of Berwick-upon-Tweed on the Scottish–English border, a town which had changed hands fourteen times in its medieval history, but which is still a place of ambiguity. There then followed two major programmes (from 1999 until 2011) around the theme of national identity, nationalism and constitutional change. They were funded by The Leverhulme Trust, without whose considerable financial support the research would not have been done and this book would never have been written. The presenting opportunity was afforded by 'devolution', the setting up of a parliament in Scotland in 1999 (and assemblies in Wales and Northern Ireland) which raised the profile of debates about national identities in these islands. As we have already mentioned, ours was not a study of 'constitutional politics', but we took advantage of the opportunity to 'get at' national identities in Scotland and in England. (We had neither the resources, nor the expertise, to study Wales and Northern Ireland.) The first programme involved a connected series of studies in which we worked with social psychologists, political scientists and social anthropologists; the second was an extended and novel series of surveys investigating aspects of 'national identity' quantitatively.

When it came to writing up this account of twenty years' research on national identity, we were confronted with our own equivalent of the 'elephant in the room', namely the referendum on Scottish independence held on 18 September 2014. This Introduction is being written after the referendum; the rest of the book went to press in late July 2014. While we touch on the issue here and there, let us repeat just once more that this book has relatively little to say about constitutional change. It follows that it is emphatically not a book about that referendum, even less the story of how it came about in the first place, nor indeed, its possible consequences.[1] Nevertheless, now that the people of Scotland have decided, it would be remiss to make no reference in this Introduction to the outcome and any impact it may have on national identity. Writing this immediately after the referendum, it

---

[1] For a perceptive assessment of the context and outcomes of the referendum, see Paterson (2015).

is already clear that although Scotland will not become an independent country in the next few years, the constitutional structure of the United Kingdom will change in ways which at present are neither clear nor predictable.

We mentioned earlier that it is not always appreciated that the connection between national identity and people's political and constitutional preferences is considerably looser than one might expect. How people 'do' national identity is a complex process and is the central concern of this book; it affects but by no means determines their constitutional preferences. Nor did the association between national identity and constitutional change alter to any great extent as the referendum approached. The latest relevant survey data (from the 2014 Scottish Social Attitudes Survey) show that while there are small variations, the relationship between national identity as usually measured and support for independence has remained broadly unchanged across the annual surveys between 2011 and 2014.

The obvious next question is whether the outcome of the referendum will change national identity in England or Scotland in the future. We could reasonably take refuge in the apocryphal reply of Zhou Enlai when asked what impact the French Revolution in the eighteenth century had on the Chinese one in the twentieth: that it is too soon to tell. At one level that is undoubtedly true. We are tempted by the usual academic mantra that more research is required and it will hopefully be done over the next few years. In its absence all we can do is make our own cautious predictions, and in the final chapters of the book, completed before the referendum, we discuss both the future of Britishness and of national identity. Writing this Introduction in its immediate aftermath, we anticipate that in Scotland not a great deal will change, though there may be a firming-up of the 'political' as opposed to the 'cultural' aspects of national identity, at least in the short term, influenced perhaps by the disputes over increased devolution which at the time of writing appear both inevitable and likely to be divisive. The future in England is rather harder to call. The promises of further devolution to Scotland which emerged in the closing stages of the referendum campaign were followed immediately afterwards by promises from the prime minister, David Cameron, of constitutional changes throughout these islands, most controversially in England where it is already clear that political consensus will be difficult, maybe even impossible to achieve. Even so, we do not anticipate rapid and dramatic change in national identity because politics and national identity are not strongly associated. Thus conflict over, for instance, how 'English' matters are handled at Westminster, or how Scotland, and the other devolved territories, are funded may well be accompanied by appeals to Englishness or Britishness without actually changing how people feel about those identities.

We incline to the view that national identity in England and Scotland will change slowly, if at all, over the next decade. Events may cause temporary shifts in national identity, though these have been noticeable by their absence in the last twenty years. We are, however, sociologists and not soothsayers. It is important to bear in mind, then, that the future may not resemble the past, and that inevitably ours is a book about the recent past, though that past is, as always, the best guide to possible futures.

1 | *Thinking about national identity*

What is it that is so interesting about national identity that it has intrigued us and dominated our research for some twenty years? Given what has already been published about nations, nationalism and national identity, why inflict yet another volume on readers? One reason is that a lot has been written about nationalism, a fair bit about 'nations', but far less about national identity. Consider this statement by the political philosopher Margaret Moore (2006: 98), in a handbook on nations and nationalism:

It would not be devastatingly dislocational, in a cultural sense, to leave Canada and live in the United States, or to leave Scotland to live in England, and would not involve the traditional costs involved in learning a new language or new symbolic repertoires. But it may be profoundly difficult for the Scot to think of herself as an Englishwoman, or the Canadian to think of himself as an American.

It seems reasonable to accept Moore's premise that it would be fairly straightforward to adjust to the new way of life, but is it more debatable that it would be 'profoundly difficult' to acquire a new 'identity'? One might of course believe that what people choose to call themselves is neither here nor there, that national identity does not really matter. We think that it does matter, and that Moore is right in her surmise. It may well be that in the two examples she gives the migrant's behaviour does not actually change very much, because so much of the new life and its meanings are familiar. How people identify themselves in national terms is another matter entirely. We know, for example, from interviews with Scots who migrate to England, and English-born people who come to live in Scotland, that their country of origin, where they were born and brought up, confers on them a powerful sense of who they are. Sometimes they did not fully appreciate just how much national identity meant to them until they went to live in another country. The sceptic may retort that this is simply sentiment, a hangover of early socialisation, and that it is a residue with little behavioural or even attitudinal force. This view is hard to defend in the light of the evidence.

In this chapter, we will ask: What is sociologically interesting about national identity? This then involves a further series of questions. Does sociology have a particular 'take' on national identity, compared, let us say, with politics, or anthropology or psychology and is it amenable to empirical enquiry as opposed to theoretical discourse? If Michael Billig (1995) is correct when he says that national identity is 'banal' in the sense that it is taken for granted, how can we ever discover what people mean by it as they go about their daily lives? Possibly national identity simply involves *not* being something? In other words, it is less about *what* you are than *who* you are not? Is national identity simply a label, an empty box, which defines what is outside, rather than its contents?

In this book, we will try to show that national identity really matters; not simply as a badge which we are given at birth and which we carry throughout our lives, like it or not. After all, none of us chose where we were born, and it is highly likely that our parents did not give it much serious thought either. There are a few people who feel so strongly about their national identity that they go to great lengths to have their babies in the 'homeland', but they seem the exception. There are also instances of people born in the 'wrong place', because they were visiting somewhere and events caught up with them, or because the appropriate hospital lay on the 'other' side of the border. When we did research in the English town of Berwick-upon-Tweed, which lies a few miles south of the England–Scotland border, we came across people who, usually as a result of medical complications, were taken to Scottish hospitals for their birth. That hardly seems significant, although when we spoke to them, they spoke of having their legs pulled by family and friends because they were 'Scots', at least by birth. It may be something of a joke, but jokes often reveal an underlying and sometimes uncomfortable truth.

Let us return to Margaret Moore's example. The Canadian may make a living south of the 42nd parallel, but continue to think of herself as Canadian, seeking out cultural news from back home, even celebrating differences as and when they arise, and possibly being a bit irritated when he is taken for an American. Such Canadians may,

for example, delight if someone can spot the differences between
Canadian and American accents. They may find that when travelling
in Europe, being 'not-American' but Canadian attracts less hostility.
Students (even Americans) have told us that, when studying in the UK,
sewing a Maple Leaf flag on your backpack improves your reception
and even guarantees you more hitch-hiking lifts on the road. Similarly,
many Scots have discovered that showing a SCO or ALBA sticker on
their car leads to less aggression from French drivers than the legal GB
sticker. Other people may regard national identity as a simple synonym of *citizenship*, reflected in having a passport. Though the US government does not really approve, American citizens can hold passports
of other countries, but still require a US one to leave and enter the
country. The passport may then be taken as a symbol of 'national identity' *tout court*, but it is by no means the same thing, even when the
term 'nationality' is thrown into the equation. What, for example, do
you write on the hotel register when it asks for your nationality? Do
you treat it as equivalent to citizenship, or do you purposefully write
down what *you* see as your 'national identity'?

If you think of yourself as a Scot, you are more likely to write down
'Scottish', although if the register specifies 'passport', you'll probably
answer 'British', just to be on the safe side. And even if you think of
yourself as 'English' you may put down 'British' because you think
they are synonyms, though there does seem to be a rising awareness
that 'English' is simply one, though numerically dominant, variant of
British. To return to our Canadian example: if you think of yourself
as Québecois you may have something of a dilemma. Quebec may be
your homeland, and you may either think of it as part of Canada, even
a very important part and say you are both Canadian and Québecois;
or you may, in the parlance, say you are Québecois and not Canadian,
which has its parallels in those who say they are Scottish and not
British. To be sure, you can always say that because you have a
Canadian or British passport whether you like it or not, that's what
you have to be, but if you had a choice, you'd put down Québecois
or Scottish.[1] In truth, it may not matter very much, because it is rare
that hotel registers have the force of law. It is not a matter of life or

[1] We included such a question in the 2003 Scottish Social Attitudes Survey, and
found that only 25 per cent of Scots would opt for a British passport if they
were given the choice, whereas two-thirds would choose a Scottish one.

death. It does, though, give us an inkling that national identity is not a straightforward matter.

Could it be of real significance? There are many examples where your life may depend on how you describe yourself: Tamil or Sri Lankan; Tutsi or Rwandan; Tibetan or Chinese. We are fortunate in this corner of Europe that our life chances, indeed our life prospects, do not depend overmuch on how we choose to describe our national identity. Yet there are apparent anomalies. Many, perhaps most, states contain different 'nationalities', groups with distinct national identities: the UK (Scots, Welsh, (Northern) Irish); Spain (Catalans, Basques, Galicians); Italy (Sardinians); France (Bretons, Corsicans); Belgium (Flemish and Walloons). When we look closely, there are few states which fit the bill of containing single nationalities; which are genuine nation-states.[2] Here, then, we uncover a puzzle: what exactly is meant by national identity if it is not simply – or even – being a citizen of a state and bearing its passport?

National identity: the weakest link?

Surely, the reader might ask, hasn't this puzzle been solved already? Can't we find the answer in the existing literature? The odd thing is that, as we said earlier, despite everything written about nations and nationalism, national identity is the underdeveloped offspring of these muscular parents. Let us take two examples, the first by the doyen of nationalism studies, Anthony Smith, who wrote a book called *National Identity* back in 1991. He begins by pointing out that 'nationalism, the ideology and the movement, must be closely related to national identity, a multi-dimensional concept, and extended to include a specific language, sentiment and symbolism' (1991: vii). Smith goes on to define the fundamental features of national identity as: an historic territory or homeland; common myths and historical memories; a common, mass public culture; common legal rights and duties; and a common economy with territorial mobility for its members (1991: 14). He says 'a nation [sic] can therefore be defined as a named human population sharing a historic territory, common myths and

---

[2] Possibly Iceland and the Scandinavian countries qualify but the Saami pose a query; Ireland is of course partitioned into North and South, and only the latter is a separate state.

historical memories, a mass, public culture, a common economy and common legal rights and duties for all members' (1991: 14). In other words, he shifts ground to equate 'nation' with 'national identity'. To be sure, Smith does not equate national identity with 'citizenship'; he appreciates that the nation need not be the state, but his account does not make it clear how one distinguishes nation from national identity or indeed whether they are really distinct. He comments: 'a sense of national identity provides a powerful means of defining and locating individual selves in the world, through the prism of the collective personality and its distinctive culture' (1991: 17). National identity is a mechanism for giving people a sense of individual and collective worth, without which they cannot function. He spells this out at length more explicitly in his concluding chapter:

The primary function of national identity is to provide a strong 'community of history and destiny' to save people from personal oblivion and restore collective faith. To identify with the nation is to identify with more than a cause or a collectivity. It is to be offered personal renewal and dignity in and through national regeneration. It is to become part of a political 'super-family' that will restore to each of its constituent families their birthright and their former noble status, where now each is deprived of power and held in contempt. (1991: 161)

While perhaps he overstates the case, Smith's view, plainly, is that national identity is the means whereby people solve the need to belong, without which they suffer alienation and atomisation. It is a psychosocial mechanism which everyone needs if they are to function as members of society. National identity sutures people into the national community and gives them meaning and purpose beyond what they themselves can generate. This is the case, it seems, whether they like it or not, or whether they believe that their national identity is actually important to them. This is, fundamentally, a functionalist explanation; individuals need national identity to function as social beings, and the nation (possibly even the state) needs their identification with its symbols and institutions in order to exist. Like any functionalist explanation, however, it raises critical questions. What if you don't identify with the nation? How do we explain that many people seem to dissent, or identify with an alternative nation? If, in Benedict Anderson's celebrated term, the nation is an 'imagined community'

(1996), how much of your own imagination can you employ, or must you simply accept what is on offer? The crux of the matter is to question how much active input, if any, people can have. Anderson himself has very little to say about national identity as such. His is an historical-culturalist account in which the actual task of 'imagining' is talked about very little, if at all. One senses that people take what is on offer, that they buy passively and willy-nilly into the nation as 'imagined', limited, sovereign and as community.

The obverse of this approach has been to try and shift the focus onto what Fox and Miller-Idriss have called 'everyday nationhood' (Fox and Miller-Idriss, 2008a). In their words: 'The nation ... is not simply the product of macro-structural forces; it is simultaneously the practical accomplishment of ordinary people engaging in routine activities' (Fox and Miller-Idriss, 2008a: 537). Their approach is to focus on four ways in which 'nationhood' is produced and reproduced in everyday life: the ways in which the nation as a discursive construct is constituted and legitimated; how nationhood frames the choices which people make; the everyday meanings of national symbols; and how 'ordinary people' 'consume the nation'. Critics such as Anthony Smith argue that the study of 'everyday nationhood' which Fox and Miller-Idriss propound has serious limitations, which 'stem from ... an underlying rejection of the causal-historical methodology common to previous scholarship in the field' (Smith, 2008: 567). This is a revealing insight into Smith's approach which rejects any idea that nation and nationalism can be studied in the here and now separated from the historical record. Furthermore he claims that to assume an undifferentiated 'ordinary people' at the heart of the enterprise is unjustified. Smith prefers to distinguish between 'elites' and diverse 'non-elites' ('castes, classes, ethnic communities', Smith, 2008: 565).

Our sympathies, as might be expected, are more with Fox and Miller-Idriss and their concern with 'everyday nationhood', although we too cannot accept the idea of an undifferentiated 'ordinary people'. However, the fact that they construe the issue as one of 'nationhood' is revealing. It is unclear to what extent their concept is identical with 'national identity', at least in the sense in which we use the term, that is as personal, generated and sustained in everyday social interaction. In their reply to Smith, they restate their aim to refocus attention on 'the everyday meaning and salience of the *nation* [our emphasis] in the world today' (2008b: 575).

Indeed, one might go so far as to say that scholars of 'nations and nationalism' have rather little to say about national identity per se. There is Ernest Gellner's celebrated aphorism: 'Nationalism is not the awakening of nations to self-consciousness: it invents nations where they do not exist' (1964: 169). In other words, the political ideology of self-determination – nationalism – creates the conditions for imagining the nation into existence. There is, of course, a large literature taking issue with Gellner's assertion,[3] but, whether it is accepted or rejected, it is essentially an argument about the relationship between nationalism and nation; from our perspective national identity as such is the presumed but unexamined hinge between them, turning the political demands into national substance, or vice versa. Gellner seemed uninterested in national identity, arguing that modern societies are characterised by high social mobility, a literate culture – 'every man is a clerc' (Gellner, 1973: 10) – and therefore there arises a pressing need for binding people culturally and socially to the nation. People have to belong, to satisfy the needs both of society and themselves, and have little option but so to do.

Scholars of nations and nationalism are not interested in national identity, except insofar as it is a by-product of connecting these elements together. Perhaps it is deemed to be a 'psychological' issue, whereby adjustment to what is expected of the person, the 'national', is the result of (early) socialisation, and is laid down and largely accepted. Where people dissent or are inadequately socialised into what is required of them, this is regarded as abnormal, a problem on and of the fringes, for example territories such as Quebec or Scotland. Such places have had the cultural wherewithal to resist incorporation into the nation's (more correctly the state's) way of doing things. It is as if one is trying to account for deviant cases, which perhaps helps to explain why so much political attention is given to making people 'citizens'.

In discussing 'national identity', it is the adjective 'national' which seems to interest scholars more the noun 'identity'. Take David Miller's comment on national identity in his book *On Nationality*: 'To understand what we mean when we talk of someone's having a national identity, we must first get clear about what *nations* [our emphasis] are' (1995: 17). Here we have a nice example of giving primacy

---

[3] See, for example, the essays in Hall (1998); and Malesevic and Haugaard (2007).

to the familiar concept, the nation. What then of *identity*? Miller observes: 'the attitudes and beliefs that constitute nationality[4] [sic] are very often hidden away in the deeper recesses of the mind, brought to full consciousness only by some dramatic event' (1995: 18). We can only respond: 'well, yes and no', given the truth as we see it of Willie McIlvanney's likening of national identity to an insurance policy which we have somewhere but are unsure of the small print. So Miller is correct, in our view, when he says that most people most of the time do not pay much attention to their national identity; it only becomes explicit in an emergency, or as he says, in the context of 'some dramatic event'. In passing, we note Billig's similar observation that nationalism is 'banal', not in the sense of being trivial, but basic or taken for granted. Where, however, Miller seems to us on less sure ground is to locate national identity 'in the deeper recesses of the mind', as if it belongs to the sub- or un-conscious ('being brought to full consciousness' etc.), and therefore almost inaccessible to social enquiry. As a political philosopher, Miller seems to have little truck with empirical research methods for getting at national identity. He observes:

Simple empiricism isn't going to settle the issue, not even empiricism of the kind that surveys people's beliefs about their place in the world. You cannot resolve the issue of Scottish nationhood by asking a representative sample of Scots, 'Do you see yourself as belonging to a distinct Scottish nation?' (Miller, 1995: 18)

Leaving aside the fact that few survey researchers, if any, would ever ask such a (mis)leading question,[5] Miller seems to dismiss all empirical enquiry.[6]

---

[4] 'Nationality' in our view is too frequently taken in the outside world as a synonym for citizenship for it to mean 'national identity'.

[5] Miller's question conflates two separate issues: 'do you think there is such a thing as a distinct Scottish nation?' and 'do you feel you belong to it?' Answering 'no' could be taken to mean: 'yes, there is such a thing, and no, I don't belong to it', as well as, 'no, there is no such thing, and I can't belong to something that does not exist'. In many years of designing survey questions on national identity, we would consider such a question incompetent.

[6] A symposium on Miller's book, *On Nationality*, was edited by Brendan O'Leary, and published in *Nations and Nationalism*, 1996. The contributions by O'Leary, Kellas and Moore are the most relevant to our book, and Kellas's in particular expresses reservations very similar to ours.

Miller's observations are useful for us because we have spent almost two decades using methods of empirical enquiry of various sorts to try and get at what people mean by national identity. We have employed intensive interviews, ethnography and social surveys, as well as working with other social scientists such as social psychologists and social anthropologists who employ experimental methods and ethnographic fieldwork respectively. Indeed, one of the purposes of this book is to argue that methods of empirical enquiry, used in the round and in complement with each other, are perfectly good ways of getting at the complexities of national identity, and that empirical enquiry is fundamental to escaping the sense that national identity is either unknowable or merely the adjunct of a concern with 'nation' and 'nationalism'.

While Miller argues that national identity (or 'nationality' as he calls it) is largely unknowable because it is usually implicit or at least below consciousness, he does acknowledge that it is, in Renan's celebrated term, a 'daily plebiscite', requiring active (hence, 'daily') affirmation that people share beliefs that they belong together (Renan, 1990 [1882]). Miller comments: 'One observes the [national] identity, weighs it against other aspects of personal identity, and so forth. There is no predetermined outcome of this process' (Miller, 1995: 44). In other words, he acknowledges that national identity is but one social identity among many (social class, gender, ethnicity, age and so on), and that individuals are active in shaping who they wish to be according to circumstances. He then puts forward an extreme view and rejects what he calls the 'radical chooser' view that the actor is supreme and can be whatever they like. Instead, he says, we balance competing demands upon us and establish our own scale of priorities between different values, thus creating 'our own distinct identity' (1995: 45). There is also the question of substance: 'Recognising one's French identity still leaves a great deal open as to the *kind* of Frenchman or Frenchwoman one is going to be' (1995: 45). One is left, then, with the sense that while Miller seems to think empirical social scientific enquiry cannot get at national identity, or even, social identity more generally, he recognises that there are some important issues to be resolved as to how people 'do' identity for themselves within particular contexts. What is unclear is how these matters could ever be resolved if they are inaccessible to empirical enquiry. We reject Miller's position partly on conceptual grounds and partly pragmatically because we believe we have successfully employed a wide range of methods to examine precisely that issue – how people 'do' identity within particular contexts.

In 1978, the political scientist Bill Mackenzie published what he described as a murder mystery. 'The victim was the word "identity", an ancient word, which once had a certain dignity. It was first harnessed to a dangerous topic in social science, that of national character, and was driven out of its wits by over-use' (Mackenzie, 1978). His specific concern was with what he called 'political identity', and he asked, like the prophet Ezekiel (37, 3), 'can these bones live?' As sociologists, we are struck by the lack of attention given to 'national identity' in the literature on social identity. Identities of social class, gender, age, ethnicity (about as close as it gets to national identity) are deemed, or asserted, to be more significant to people in their everyday lives: so much for Renan's daily plebiscite. As we have already indicated, where it *is* discussed, it is usually in the context of citizenship. For example, in the political discussion about 'Britishness' in the first decade of twenty-first-century Britain, government concerns seemed mainly about assimilation and multiculturalism, of adapting ceremonial from the USA to make immigrants 'feel British', and insisting that they should take classes in citizenship. So we have something of a double veil when it comes to national identity. On the one hand, there is very little attention given to national identity per se in the abundant literature on nations and nationalism. On the other, writing on social identities does not afford it much of a place either.

'National identity' is not simply the property of social scientists. In a usefully comprehensive review, Peter Mandler argues that historians

can offer social scientists a lot of useful information about the symbolic world that is available to social actors out of which they construct their group identifications. Social scientists understand much better the process by which group identification takes place than they do what determines the content of group identifications or even the salience of a particular group identification in a given situation. (Mandler, 2006: 276)

While observing that 'identity' is an important but elusive quality, and 'national identity' even more so, he comments that 'what goes on in people's heads is very complicated and difficult for historians to pin down' (Mandler, 2006: 281).

It is significant that it was a social psychologist, Michael Billig, who was prepared to talk about 'national *identity*' in his book *Banal Nationalism* (1995), and there is a rich tradition from within that discipline, reflected, for example, in Reicher and Hopkins' book *Self and Nation* (2001). Billig has been taken to task for overlooking the complexity of 'national life', especially in Britain, and for lacking any sense of dynamism (Skey, 2009: 342). In reply, Billig points out that people (ordinary or otherwise) by no means passively receive a single ideological message, notably as it comes through the media. Thus, 'Each individual is likely to have contrary things to say, as they seek to balance the conflicting themes of common sense' (Billig, 2009: 348).

There are, however, those who would reject the whole notion of 'identity'. Writing with his colleague Frederick Cooper, the American sociologist Rogers Brubaker dismisses the concept as explaining everything and nothing, such is its lack of specification (Brubaker and Cooper, 2000). As Richard Jenkins points out, Brubaker's real attack is focused on the concept of 'group', notably 'ethnic group', which assumes too much homogeneity, boundedness, and hence, 'reality'. As Jenkins observes: 'Ethnicity for Brubaker is cognitive, a point of view of individuals, a way of seeing the world' (2008: 9). Thus, says Brubaker, identity is not a 'thing' which people have or don't have, or aspire to, and certainly cannot make people 'do' anything. People make and do identity for their own purposes and in their own ways. In the language of empirical analysis, identity is an *explanandum*, not an *explanans*, a dependent not an independent variable. However, Jenkins observes: 'It is only the definition of groups that he uses – as definitely bounded, internally more or less homogeneous and clearly differentiated from other groups of the same basic kind – that allows him to reject their reality' (2008: 12). Observes Jenkins, 'groups may be imagined, but this does not mean that they are imaginary' (2008: 12), an echo perhaps of the debate about Benedict Anderson's 'imagined community'.

We do not intend in this book to offer a thoroughgoing account of the literature of social identity. Richard Jenkins' book, now in its third edition, has a great deal to commend it, and there is much in it which complements our approach to national identity. For example, we have a lot of sympathy with his comment that we need a compromise between rejectionists like Brubaker and Cooper on the one hand, and those who, on the other, accept uncritically the ontological status of 'identity', social or national. There is as lit-tle point sweeping away notions of national identity as so much false consciousness, as there is to treat it as a fixed given, used as a 'cause' to explain variations in human behaviour. Perhaps too it would be helpful to get away from 'identity' (as a noun), implying that it is a badge which affixes to people, describing who they *are*, as it were, and treat it more as a verb, 'to identify with', which implies a more active process of doing, which varies according to context. In this way, by treating people as agents, we are better able to get at how and why people mobilise national identifications, and for what purposes; what it means, for example, to make claims both for oneself and for others, and to judge such claims in the contexts in which they are made. We approve of Jenkins' view, citing that of the social anthropologist Fredrik Barth, that ethnic (and, by implica-tion, national) identification is transactional and situationally flex-ible (Jenkins, 2008: 124).

Quite early in our research on national identity, we revisited the writing of Erving Goffman whose classic book *The Presentation of Self in Everyday Life* arose out of his fieldwork in Shetland while he was at Edinburgh University doing his PhD, and was first published in 1959. Goffman, in this and his later works, had nothing to say expli-citly about national identity, but we were taken with his approach that individuals have considerable leeway in constructing and mobilising how they presented aspects of self, and that, in the last resort, it was that 'presentation' which determined how successful or not one was in social interaction. Ours was not an interest in the micro-sociological aspects of personal interaction, but in who people thought they were in national terms, who they thought others were, and how they managed these conceptions of national identity in interaction. To paraphrase Jenkins' writing about ethnicity, national identity is a two-way street, involving a sense of *them* as well as a sense of *us* (2008: 123). It also involves thinking of oneself as *not* being someone, as much as being someone; that is, difference as well as similarity. It is not the case, we think, that people simply play out the parts allotted to them by the 'nation', still less the state; they are not puppets on a stage. Neither do they have the capacity to make it all up as they go along; Miller's 'radical choosers' again. The analogy with presentation of self is clear. Somehow, individuals are able to play the parts they want in the way they want, but within certain constraints. Here is another interesting comment by Jenkins, where he argues that claims to identities need to have some objective correlates to be accepted by the audience:

A group of people without Norwegian passports, with no discoverable historical connections to Norway, and speaking no Norwegian, cannot simply arrive at the Norwegian border and have any expectation of mounting a plausible claim to Norwegian identity or nationality. (2008: 127–8)

There have to be some cultural or social markers which form the basis of such a claim, whether in terms of citizenship, ancestry, language, or 'history', to pick out the ones Jenkins uses. Leaving aside the fact that Norway is a state and Scotland is not (they have virtually the same population size – around 5 million), we recognise the social game being played. In our work, we have looked to see how people make use of cultural markers such as place of birth, ancestry, residence, skin colour, in assessing whether or not, and in what circumstances, to make claims to national identity, and to judge the claims of others.

No such thing as society?

Before we explain in more detail how we have gone about doing research on national identity, let us make clear where we stand on how we situate individuals vis-à-vis society in conceptual terms. By placing our methodological weight on how individual people 'do' identity, we are absolutely not implying that they have a free hand so to do. We have observed already that who we are depends to a large extent on how well or badly others, our audiences of significant others, affirm or reject our claims to be who we would like to be. In discussing how Caribbean identities are negotiated, the sociologist Stuart Hall made a more general point that in a key sense, identities 'come from outside', that 'without others there is no self, there is no self-recognition' (Hall, 1995: 4).

There are many discussions of the dialectic between the individual and the social, and it is not our purpose to set these out, except insofar as they tell us something important about national identity. The legal theorist Neil MacCormick once wrote:

The truth about human individuals is that they are social products, not independent atoms capable of constituting society through a voluntary coming together. We are as much constituted by our society as it is by us. The biological facts of birth and early nourishment and the socio-psychological facts of our education and socialisation are essential to constituting us as persons. We are the persons we came to be in the social settings and contexts in which we find ourselves, and whatever sense we have of our identity and character as persons reflects our interaction with significant others in our social setting, and indeed in a more diffuse way is a reflection of our total social milieu. (MacCormick, 1999: 163)

Why quote a legal theorist, albeit clearly a sociologically sophisticated one, rather than, say, a social scientist? MacCormick made this statement in the context of developing an account, as both a legal theorist and an active (Scottish) nationalist, of 'liberal nationalism', arguing that nationalism was not inevitably 'ethnic', but on all fours with civic, liberal society. In the words of Yael Tamir, we are talking about the 'contextual individual' who combines individuality and sociability as two equally genuine and important features (1993: 33). In rejecting methodological individualism, MacCormick is arguing that we cannot have a model which separates the individual from the social such that the former precedes the latter; that in the (in)famous words of Mrs Thatcher: there is no such thing as society; there are only individuals and their families. For MacCormick and Tamir, there is no conflict between liberalism and nationalism. In MacCormick's words:

Contextual individuals may have as one among their most significant contexts some national identity. On that account, respect for national identities, and acceptance of the legitimacy of a civic-cum-personal variant of nationalism, do not conflict with liberalism. Indeed, liberalism may require this.

He goes on to say:

it is not theoretical nationalist imaginings, but the facts of political life, that give national identity a special place in the contextual definition of the contextual individual in her/his character as a political animal. (MacCormick, 1996: 565 and 566)

The fact is that that there is such a thing as society, and we constitute and create it.

What follows?

In this book, we will address and attempt to answer a series of questions which are not only of academic but of wider social and political interest. To anticipate any misunderstandings it is worth emphasising at this point that this book is not about the recent Scottish independence referendum and was almost entirely written before the referendum took place on 18 September 2014. We have since then inserted a few passing comments and a slightly more extended discussion in Chapter 9 during the final stages of production but the result makes little difference to the central discussion and arguments we advance. As we shall see later, national identity is only loosely associated with attitudes to constitutional change, which in turn are only loosely associated with political preferences. What will happen in the coming months and years is of course hard to predict but the best guide to the future is the past, and how Scots view their national identity, at least as it is conventionally measured, has changed remarkably little over the last fifteen years or so. The pattern has hardly changed since devolution. Partly this is because national identity in both England and Scotland has a strong cultural rather than political dimension. We return to these issues and others later in the book.

In this chapter, we have outlined what we see as the main issues about the concept of national identity. In the second chapter, '*Accessing national identity*', we explore how to 'get at' national identity, which is largely, though not entirely, a methodological issue; how is this to be done?

In the third chapter, '*National identity: do people care about it?*', we will explore whether it actually matters to people, and ask, if so, to whom? If 'national identity' is a concept which people use in their social lives, and see as a meaningful social category, how does it relate to other social identities they may have – social class, gender, ethnicity, age and so on?

In the fourth chapter, '*Debatable lands: national identities on the border*', we examine how people living in Berwick-upon-Tweed, a town on the English–Scottish border, make sense of their interstitial territorial status living as they do in English jurisdiction, but in a town so close to the border which historically belonged to Scotland.

National identity, like other forms of social identity, is not simply about our own senses of self, but whether or not we accept the claims of others. That is the theme of the fifth chapter, '*Claiming national identity*'. Who do people consider to be 'one of us'? Do birthplace, ancestry, accent matter more than skin colour or 'race'?

It is hard to escape the notion that national identity is really about 'politics', and it is important that we discuss the relationship. We will explore these topics in Chapter 6, '*The politics of national identity*'.

All forms of social identity, including national identity, involve 'othering': the extent to which we measure who we are in terms of who we are not. This is the theme of Chapter 7, '*The notional other: ethnicity and national identity*'.

In Chapter 8, '*A manner of speaking: the end of being British?*', we examine the question of what it means to be British these days. If the people of these islands are becoming more aware of their 'national' identities, does this have implications for the future of being British?

In the final chapter, we explore the question, '*Whither national identity?*' Rather than seeing national identity as ephemeral, even chimerical, we argue that it raises fundamental questions about the nature of the social and social order in particular. As well as reviewing how our approach to the study of national identity has evolved over two decades, we argue that the issues surrounding it have become more, not less, important.

Social identities can be considered the fulcrum around which issues of social structure and social action rotate, and without understanding national identity in particular, we fail to connect the personal and the social.

Let us begin with the fundamental question: how do we 'get at' national identity, if at all? This is the topic of the next chapter.

2 | *Accessing national identity*

If, as we argued in the previous chapter, national identity matters, that it is important to social actors, as well as being a useful concept for explaining their attitudes, actions and behaviours, how do we study it – in other words, how do we 'get at' national identity? This is not simply a matter of adopting the *best* research method, assuming such a thing ever exists, or a variety of methods, but closely connected with how we conceptualise national identity in the first place. In this chapter, we will give an account of our understanding of national identity, how we went about studying it, the methods we employed and the research designs we adopted.[1] We will discuss our findings in the chapters which follow.

There are two influential views in this regard. Firstly, there is the commonsense view, shared as we have seen by some theorists, that national identity is a fixed badge, something we all have been given, derived from our 'nationality', and there is little more to be said. Once we have it, a bit like an identity card, we are what it says on the tin. From this perspective, we can use it or ignore it, but there is nothing much we can do to change it, short of emigrating and taking another citizenship, or perhaps forging another identity. That, though, is wrongly to equate national identity with 'nationality', or citizenship. Nor is it simply a matter of where you were born. National identity is much more subtle and variable than that. It will probably be clear to the reader that living and working in Scotland has been a major influence on our thinking. 'Nationality', British citizenship, is not at all the same as 'national identity', being Scottish or indeed English. Our understanding of national identity, then, comes literally with the

---

[1] The purpose of the second half of this chapter is to discuss a variety of methodologies for studying national identity and we do not generally discuss the results. For accounts of what these studies found, see *passim*, Bechhofer and McCrone (2009b).

territory. We understand the distinction between (British) state and (Scottish) nation because it is in common currency north of the border. That, of course, carries its own limitations: we might assume, wrongly, that other nations in the state, notably England, have the same sets of understandings. One might think that people who make what Scots would regard as a category error, 'Britain equals England', must somehow fail to understand the state–nation distinction, that it is some kind of cognitive failing. It may in fact be telling us something more subtle about social and cultural processes south of the border (see Condor et al., 2006; and Condor and Fenton, 2012). In short, it's a matter of sociology, not of cognition. Be that as it may, living in a society in which state and nation are generally understood not to be the same thing does carry the risk of assuming that it must be so elsewhere, and plainly it is not.

If one error in studying national identity is to presume that it is fixed and that most people, if not all, share a fairly similar conception, a second is to reject national identity as epiphenomenal, as somehow 'unreal' in terms of how people go about their lives. It then becomes at best an analyst's term, of little relevance to people's everyday lives, and is rarely seen as one they employ to explain who they are and what they get. Think back to the comment by McIlvanney in the Introduction: most of us know vaguely that we have a national identity, but we are not sure what precisely it means. There is an important strain of academic writing about national identity which cautions against making the concept do too much analytical work. For example, Susan Condor and Jackie Abell made this important point:

'Identity' is a notoriously polyvalent term and a consideration of its use in the current literature on nationalism reveals a common tendency to elide the 'national identity' construct with pre-existing academic categories such as nation, nationality, nationalism, national character, citizenship, or imagined community. In addition, the ambiguities of the referent of the term 'identity' afford slippage between 'the nation' as an object of literary or political rhetoric and assumptions concerning the subjective self-consciousness of individual citizens. (2006: 52)

Condor and Abell are critical of the tendency to treat national identity 'as an analyst's concept rather than a participant's resource', and are keen to 'consider ways in which ordinary social actors may construct nation-ness as a matter of subjective identity' (2006: 53). Another example of seeing national identity as a 'discursive construction' is the study by Wodak and her colleagues, who argue that

The national identity of individuals who perceive themselves as belonging to a national collectivity is manifested, inter alia, in their social practices, one of which is discursive practice. The respective national identity is shaped by state, political, institutional and media and everyday social practices, and the material and social conditions which emerge as their results, to which the individual is subjected. (Wodak et al., 1999: 29)

The focus on 'ordinary' social actors is something with which we have considerable sympathy, although we are less certain about analysing national identity simply in terms of 'discourses of nationhood'. We do not read these authors as saying that national identity belongs simply to 'discourse', that it has no real meaning for actors themselves. Far from it. To anticipate somewhat: while we acknowledge that social and cultural talk in society at large, for instance in the media, is important in framing how social actors define and employ national identity, we share their view that getting people to talk about it in their own terms is the key to understanding what it means to specific individual actors.

Perhaps where we part company with our colleagues is with regard to the methods to be used. We have relied much more than they have on survey methodology alongside other forms of data collection and analysis, such as intensive interviews, and in the work we have done with others: ethnography, social psychological experiments, documentary analysis and so on. Because we are aware that some scholars are sceptical about the use of survey methods in the study of national identity (see Miller, 1995; Malesevic, 2011), it is important to stress that survey questions cannot just be dreamed up out of thin air. They have to be based on prior knowledge gained from detailed analysis of 'identity talk' in different contexts. Indeed, we have found considerable benefits in employing 'triangulation', the view that several observations of a datum using different research methods are better than one (Bechhofer and Paterson, 2000: 57). Thus it is desirable to use different methods to get at national identity, and where possible, cross-fertilise them. A simple example would be to use the ways people talk in a non-structured format in order to fashion survey questions which can be asked more generally, and enable responses to be compared more directly. For instance, if we were to find that people talked in particular ways about 'politics' when they discussed national identity, we might design survey questions to see how widely people held to this view.

This is not to advocate methodological promiscuity, as if we can pick a method off the shelf without due acknowledgement of the epistemological assumptions underpinning it. Rather, it is to recognise that what we mean by 'national identity' is the property of analysts *and* social actors alike. Like so many concepts and ideas in social science, we are dealing with what is already in the public domain. This is both a help and a hindrance. Just as 'social class' and 'gender' and 'race', to name three, are understood and used in social practice, though often not in the ways social analysts use them, so 'national identity' has various meanings in the social and cultural world. Social scientists are fairly practised at handling such conflicts of meaning, often being accused either of telling people what they already know, or, if it does not square with deeply held beliefs, that they must be wrong and don't know what they are talking about. Both, obviously, cannot be true at the same time, and are nice examples of what Paul Lazarsfeld once referred to as 'stating the obvious' (1949). We ourselves took a leaf out of his book in a paper called 'Stating the obvious: ten truths about national identity' (Bechhofer and McCrone, 2009a). In this paper we laid out ten *ex cathedra* statements which had the ring of widely accepted truth and then showed that the evidence proved them to be incorrect. The 'truths' turned out not to be so 'obvious' after all.

As we indicated in the previous chapter, our own work on national identity has drawn broadly on the sociological perspective of symbolic interactionism, focusing on the performative and presentational aspects of identity. Though neither of us in our previous work had carried out research at this micro-sociological level, we were attracted to the work of Erving Goffman, who saw social identity (as far as we know, he had nothing to say about *national* identity per se) as a tactical construction, designed to maximise player advantage. In particular, we developed his idea that identity is a tactical issue involving claims, how these are received, and how identity characteristics are attributed to actors on the basis of what the audience receives and interprets. Our research was not designed in any way as a test of Goffman's theories, but we found his approach, and the wider work of symbolic interactionists, instructive. Another example is W. I. Thomas's classic idea (2009 [1928]), also drawn from symbolic interactionism, of the 'definition of the situation' whereby regardless of truth status, people act on the basis of what they believe to be the case.

With regard to national identity, then, we hold that it cannot be defined a priori, as if it had an 'objective' character outwith and superior to the meanings implied by social actors. Just as it is not possible to uncover the objective characteristics which constitute a 'nation' (and the literature is replete with failed attempts), so seeking out the 'true' aspects of national identity – indeed, possibly any form of social identity – is a thankless task. We see social identity in general, of which national identity is one form, as something of a hinge between social structure and social action. Social structure constrains but does not determine how people behave, yet social action is not entirely a matter of free will. Identity provides a set of meanings and understandings through which people experience social structure and feel empowered to act. By adopting a Goffmanesque approach,[2] one can see that social actors might well have more influence over aspects of national identity than a top-down approach implies. They neither, then, act out a set of pre-ordained practices as required by those in authority or institutional convention; nor do they make them up *de novo* as they go along. People have considerable influence over how they choose to negotiate and mobilise 'national identity'. Using a quite different theoretical approach to our own, Tim Edensor in his study of how national identity is represented and transformed in popular culture has observed that '"national" is constituted and reproduced, contested and reaffirmed in everyday life' (2002: 20). In our work we have tended to take national identity at 'face value', not in the sense that anything goes, but that it is a matter of presentation, claim and counter-claim, that is, of 'face'; in other words, people wish to be judged on the basis of their claims. That is why we found Anthony Cohen's concept of 'personal nationalism' interesting, as it highlights the actor as a 'thinking self' (1994: 167), and not a stage puppet acting out a role dreamed up by powerful managers.

---

[2] We use 'Goffmanesque' to indicate that we are exploring national identity in the manner of Goffman, rather than according to the strict tenets of his approach.

The importance of research design

Perhaps, when it comes to studying national identity, it is the importance of research design which has been underestimated. Put most starkly, if most people take their own national identity for granted, as an unproblematic given, it is not straightforward to get at national identity in *those* circumstances. Thus, it is those people in problematic situations or places who turn out to be interesting and illuminating. They are not 'typical', but they highlight processes or experiences which tell us something about how identity operates, for whom and under what circumstances. Two examples might help to make the point. One of our earliest pieces of research focused on 'elites', one a group who owned considerable amounts of land in Scotland, and the other those who managed key cultural institutions, the 'arts elite'. Neither the landed nor the arts elites were 'typical' of the population at large, but they had considerable media salience at the time we carried out the study, which has resurfaced from time to time since, and again at the time of writing.[3] This meant that they had become practised at talking about their national identity, and how they justified owning large amounts of (Scot)land in one case, and running key national cultural institutions in the other. The point was that their high profiles raised the salience of their national identity, and made it a topic of debate quite unrelated to their experience or competence. If they had not been born and brought up in Scotland, as many had not, they had to mount a defence of owning or running something as iconic as land or culture. It was this piece of research which generated our interest in the markers of national identity, and the rules which operated around them. Thus, many in these elites could not fall back on the usual criteria ('markers') of birth and upbringing as justification, but had to make much more of living, and earning a living, in Scotland, making a contribution to the society, and crucially, deciding to make a commitment to the country. They could argue that the birthplace marker of national identity was just a fact of life outwith their control, whereas the positive action of choosing to live in the country and make a commitment to it was the key to their identity.

---

[3] See, for example, www.scotland.gov.uk/About/Review/land-reform, published in May 2014; on arts in Scotland, see www.bbc.co.uk/news/uk-scotland-19880871 (accessed 16 June 2014).

Another piece of early research followed the same logic of the unusual, and critical, case. We moved, as it were, from *people* whose national identity was publicly contested, to debatable *places*, in that we did interviews and ethnography in Berwick-upon-Tweed, a town sitting on the border between Scotland and England, and which, in its history, had changed hands between the two countries fourteen times, before becoming a jurisdictional part of England in the late fifteenth century. The point is a simple one: places on a border are often those with contested identities, even to the point of 'frontier peoples' being *über-national*.[4] Our research question was a straightforward one: how do the residents of this debatable place think of themselves in terms of national identity? In legal-jurisdictional terms, they are indubitably 'English', but we discovered that a large number of those who resided in 'old' Berwick, the part within the walls of the town on the north bank of the river Tweed, were more likely than residents of the borough on the south bank to say that they were either Scots or 'Berwickers'.[5] We could not be sure, however, that any ambivalence about identity was not simply a reflection of being peripheral to the main centres of power. To that end, we carried out similar work on two communities on either side of Berwick; the town of Alnwick, and some small surrounding villages, to the south, and Eyemouth to the north. If identity were simply an expression of peripherality, then one might have expected the residents of these places to talk in similar if perhaps more muted terms as the people of Berwick. Rather, we found that the people of Eyemouth saw themselves as unequivocally Scottish, and those in Alnwick and local villages south of the border as English. In other words, we could be pretty sure that it was Berwick's geographical, historical and cultural status that made it a place of ambiguity in national identity terms, something we will discuss further in Chapter 4.

These two examples, of 'contested' people and places, serve to make the point that good research design can potentially make processes

---

[4] Where borders and frontiers are disputed, national identity tends to become more salient. See, for example, in the former Yugoslavia, Malcolm (1994) and Judah (1997). On Northern Ireland, see O'Dowd and Wilson (1996).
[5] In other words, they opted for 'local' identity as a way of avoiding having to choose one or other 'national' identit. In the words of a local ditty: 'They say there is England and Scotland indeed. There's England and Scotland and Berwick on Tweed.'

of national identification more transparent. Both involve comparison, both in terms of internal variation (landed vs arts elites; Berwick vs Eyemouth/Alnwick), but also, by implication, with the rest of the population. The research locales are meaningful and critically chosen (Bechhofer and Paterson, 2000), and thus yield insights potentially generalisable to other people and places in the mainstream.

Markers and rules

This early work on 'contested' people and places alerted us to what we came to describe as the 'markers and rules' of national identity. In other words, what are the criteria people use when they claim national identity, and how do they judge the claims of others? Contenders include place of birth, accent, ancestry and place of residence. For example, many members of our landed and arts elites had not been born in Scotland, and so had to have recourse to other markers when people judged them not to be Scottish and queried their right to own the land or to direct cultural institutions. How could people recognise these markers? Patently, they did not have access to birth certificates, but made judgements on the basis of proxy markers such as accent. Thus, if someone spoke with an 'English' accent, the assumption was that they were born in England, although someone born in Scotland but brought up from an early age south of the border could be 'misread' as English. This work required us to be more systematic in our assessment of national identity markers. Most obviously, some markers were more transparent than others; and some were more fluid or fixed than others. Thus, we might treat accent as more or less fixed (unless the person is an unusually good voice actor), and as readily accessible. We then 'read off' the messages it gives out. Other markers are more fluid such as place of residence, for example, one which is by definition visible to others. On the other hand, place of birth and upbringing, as well as ancestry, are more opaque; we have to take people's word for it, by and large.

Then there are problematic and sensitive markers such as skin colour, what Michael Banton (1999) has called a phenotypic marker, around which there is considerable controversy. Can a non-white person be 'one of us'? And on what grounds can they so claim, or have their claims denied? We can, of course, ask non-white people how they would describe themselves, but this is far more an issue about how the white 'majority' view *them*, and under what conditions this 'majority' are prepared to accept claims to national identity. Can a non-white person be British, English, Scottish? This is tricky territory for many people, running close to racism, and something to which parties of the extreme right are happy to appeal. We know from official statistics (McCrone, 2001: 171) that non-white people living in England are more likely to emphasise their citizenship/nationality and say they are 'British' (as opposed to 'English'), than are their Scottish equivalents. This suggests the possibility that 'English' carries more of an ethnic than a civic meaning, but that is a question we will analyse later.

Our early interest in 'markers' of national identity, the criteria people use to make judgements about their own national identity and that of others, emerged from our work on elites in large part because we were asking respondents to talk through the challenges they faced either as owners of land or managers of cultural institutions. Further, we were able to identify 'rules' which seemed to govern how markers were used. By rules we do not mean what one might find in a formal rulebook, still less laws, but a set of 'rules of the game' usually implicit in social interactions and understandings. In other words, we know how to 'behave' in social settings because we have learned the norms as to what is acceptable and unacceptable. Coming new to a situation puts people at a disadvantage unless and until they pick up these norms. Let us take a specific example. By and large, we know from survey work, as well as what people told us, or implied, in interviews[6] that the taken-for-granted criterion for being 'national' is where you were born and brought up. Thus, a very straightforward rule of national identification would simply be to say that you are a Ruritanian because you were born in Ruritania. For those not born there, that is a challenge, but they may respond by pointing out that no-one can determine where they were born, but where they want to live is up to them. We assert our commitment to a place by choosing to live there. Or they might try an alternative tack and stress that they have important ancestral links, that they are 'national' because of their blood links. This can be tricky, because claims based on blood, ever

---

[6] In the Scottish Election Study of 1997, 82 per cent said that birthplace is a very or fairly important criterion of Scottish national identity, followed closely by ancestry (73 per cent) and residence (65 per cent).

since the experience of fascism in Europe, are challenging.[7] They are often used by the racist Right to justify their politics, by claiming that even people who happen to be born in a country may not have the acceptable bloodlines which make them 'properly' national.

The point is that how people choose to play their hand is not determined a priori. The game metaphor is useful. All of us are dealt cards (national identity criteria, in this case) which we play with greater or lesser skill to achieve our preferred ends. We may even choose not to play what appears to be our strongest card because we have other games to play. Let us take an example from recent British politics, with regard to three leading players: Tony Blair, Gordon Brown and David Cameron.[8] Of the three, Gordon Brown seemed to have the strongest suit when it comes to being Scottish: born and brought up in Scotland, of indubitably Scottish parents, and with a Scottish accent. By and large, being Scottish is not a card he chose to play explicitly because he had another, a Unionist, agenda – being British. It goes without saying, of course, that he was regarded as Scottish,[9] and not only had no need to play the card, but believed he should offset it by recourse to the wider – British – claim. If place of birth inevitably conferred national identity, then Tony Blair would also be 'Scottish', but as far as we know, he never played that card, at least in public. We cannot really tell whether he regards himself as Scottish, but he too seems to have had a wider agenda relating to making a broader appeal to what is conventionally construed as 'middle England'. So we have the situation that two British prime ministers since 1997 have not forefronted their Scottish credentials because they wanted to play a wider political game. The third of our political leaders, David Cameron, is, as far as one can tell, never taken for Scottish, nor has he explicitly claimed to be such. He wasn't born there; he doesn't 'sound Scottish', but he does

---

[7] The political potency and extreme resonances of 'blood and soil' (*Blut und Boden*) were reflected during the Scottish referendum campaign (www.independent.co.uk/news/uk/politics/scottish-independence-alistair-darling-accused-of-comparing-yes-voters-to-nazis-and-says-alex-salmond-is-behaving-like-kim-jongil-9490886.html, accessed 16 June 2014).

[8] During the 2010 UK general election campaign, some right-wing newspapers accused the Liberal Democrat leader Nick Clegg of not being 'British' enough because he had a Spanish wife, a Russian father and a Dutch mother, even though Clegg was born and brought up in the UK.

[9] And post-devolution, it is often played against him by political opponents.

carry a 'Scottish' surname, Cameron, and he does indeed have Scottish roots. In the Scottish referendum campaign, he chose to play this card. In a speech in February 2014, he said: 'My surname goes back to the West Highlands and, by the way, I am as proud of my Scottish heritage as I am of my English heritage. The name Cameron might mean "crooked nose" but the clan motto is "Let us unite" and that is exactly what we in these islands have done' (www.bbc.co.uk/news/uk-politics-26082372, accessed 19 February 2014).

Hitherto, Cameron saw little need to play the 'Scottish' card, and in truth, he has no political need to do so; indeed, he might see it not only as counterproductive ('not *another* Scot as prime minister'), but his claim might be disparaged north of the border. However, Cameron is not claiming to be Scottish, but using an ancestral claim to make a point about the diverse 'ethnic' origins of the British ('I am a classic case. My father's father was a Cameron, my mother's mother was a Llewellyn. I was born and have always lived in England.'). If, on the other hand, he made the 'strong' case for being Scottish, and turned up to an event wearing a kilt of Cameron tartan, for example, he would probably be the object of some derision,[10] not least among Scots.

Let us give another example of the point that one cannot determine a priori how people will play their cards. As we stated earlier, non-white people in Scotland of Asian origin are more likely to claim to be Scottish than their counterparts in England are to make English claims (McCrone, 2001: 171). Much depends on the cultural-political context of such claims. A shrinking population, campaigns by successive Scottish governments to encourage in-migration, to tackle racism as 'un-Scottish', to reinforce a belief, which has elements of myth, that Scots are a friendly and welcoming people, and, crucially, a political consensus among parties at Holyrood, coupled with the electoral weakness in Scotland of racist (BNP) and chauvinistic (UKIP) parties, all combine to encourage such claims. As we shall discuss later, it is in reality the case that 'native' populations in England and Scotland differ very little as regards their willingness or otherwise to accept such claims from non-white people to be 'nationals'. The key point is that the political-cultural contexts in the two countries are

---

[10] Strongly nationalist Scots probably found the speech somewhat absurd but of course Cameron was not appealing to them, a lost cause, but to the undecided voters in the middle.

different and help to shape people's willingness to play the national identity game.

So far, we have tried to show that not only is national identity a fairly implicit social characteristic, but that it involves quite complex sets of markers which are employed by people with greater or lesser skill as and when required. Identity is not a given, nor on the other hand is it so flexible as to be meaningless. We might accept that national identity is most of the time 'unimportant' in people's lives, except when it is activated or mobilised by politicians or other cultural entrepreneurs, such as the media. In other words, identity talk involves communicative practices; it is not immutable. As the social psychologist Margie Wetherell observed:

identity needs to be 'done' over and over. What 'it' is and who 'we' are escapes, is ineffable, and needs narrating, re-working, and must be continually brought 'to life' again and again. It is only in certain limited contexts such as autobiographies, *Hello* magazines and in immigration halls (important as these things are) that identities become finalized and accomplished once and for all time (and usually not even there). (Wetherell, 2009: 4)

Are surveys good enough?

If national identity is such a subtle, nuanced and transitory phenomenon, how can social scientists best study it? Social surveys, surely, are too crude to get at the subtleties involved? That is to misunderstand the nature of surveys. Questions are not necessarily completely structured but have varying degrees of openness. Above all, they ought to be employed alongside other methods, as complements rather than competitors. As we have pointed out, we have used survey questions in conjunction with intensive interviews involving talking to people at considerable length (between one and two hours) about national identity. Indeed, our early qualitative work on elites generated issues and questions which eventually, much later, lent themselves to a survey format. Let us, then, review how surveys have handled national identity.[11] The gold standard of British social surveys, the British Social Attitudes Survey which began in 1983, has asked two related questions

---

[11] For a review of structured questionnaires, see Bechhofer and Paterson, 2000: ch. 6.

in Scotland and England. First, multiple choice: showing a list of identities, British, Scottish etc. and allowing respondents to choose more than
one identity so that they can be both British and English, for example;
and then as a follow-up, asking which they would choose if restricted to
just one choice. More recently, it has adopted what has become known
as the Moreno question, which strictly should be the Juan Linz question, originally designed for use in public opinion surveys in Catalonia
and Spain using a five-point Likert scale. Adapted for use in Scotland by
Luis Moreno (Moreno, 2006), it has been used north of the border since
1986, extended to England in 1997 and to Wales in 2001. It seemed
to us a more subtle and sociologically preferable measure which allows
respondents to choose from a continuum of options, for instance in the
Scottish case: Scottish not British, more Scottish than British, equally
Scottish and British, more British than Scottish, and British not Scottish.
It has the merit over earlier measures of allowing people to combine
'national' and 'state' identities, without forcing a choice between one
or the other, or implying that if a respondent chooses two options (say,
British and Scottish), they are treating each as equally important. It is
worth noting at this point that it should not be interpreted as reflecting
the *strength* with which views are held. To date, the aggregate results have
been consistent over a considerable time period, with around two-thirds
of people in Scotland giving priority to being Scottish (either 'only' or
'mainly'), and only around one in twenty to being British. The only piece
of longitudinal research to date, using data from a cohort surveyed in
1997 and again in 1999, suggests that people are also fairly consistent
over time, with around three-quarters giving the same category response
on both occasions and very few shifting by more than one category. It
does then seem reasonable to argue that survey data of this kind are good
at capturing reliable snapshot pictures of how large numbers of people
describe themselves. The suspicion remains, however, that the measure
has problems of validity, that it is not measuring what it purports to. For
example, it could be claimed that people in England in particular simply
use a different label to describe the same thing, saying they are English
rather than they are British, but at a deeper level meaning the same thing
by the two terms,[12] perhaps being swayed by changing usage in everyday
currency.

---

[12] This is the view of Anthony Barnett (1997 and 2013) which we will elaborate
in the next chapter.

We cannot assume, either, that people intend to convey exactly the same thing by using the same identity category. Our qualitative interviews indicate a wide variation in responses among people opting for the same category, some simply asserting their bald choice ('this is who I am'), and others talking through how they came to opt for one rather than the other. Here are two contrasting examples:[13]

R1: I would say that I'm a wee bit more Scottish than British but I still feel British ... Because I am Scottish first and foremost, you know.

And another:

R2: No. 1 [*Scottish not British*] is out, 'cos I do think myself as British ... No. 3 [*Equally Scottish and British*] is out. I was looking at that one but I said I was Scottish first, and I stand by that. I live here, if I'd lived in England all my life I would probably have said '*More English than British*'. But if I chose '*Equally Scottish and British*' it would mean I would accept a British football team which I wouldn't, I want a Scottish one.

What we get from the latter respondent is some kind of explanation for their choice, in contrast with a fairly matter-of-fact and non-elaborated assertion from the former. Plainly, this makes any kind of systematic comparison between such responses difficult. Our qualitative interview data do, however, suggest that 'Moreno choices' are meaningful. Based on how people discussed their reasons for choosing a category, we developed for each national identity response a series of statements expressing these reasons. For each we asked them, once more on a five-point Likert scale, how strongly they agreed or disagreed. The reasons ranged from attitudes to Empire, to history, traditions and culture, to national institutions, social values and to devolution. Examples are 'I feel uncomfortable about the idea of being British because I want to distance myself from the British Empire and all it stood for' and 'I identify with [English/Scottish] history, traditions and culture' (Bechhofer and McCrone, 2010). By using factor analysis we can then identify the underlying dimensions and gain some

[13] Quotes from respondents which occur as a sequence in close proximity are numbered to distinguish one from the other. Note that a respondent, R1 say, at one place in a chapter is not the same person as R1 in a later sequence in the same chapter or in another chapter unless specifically so stated.

idea of the structure lying behind people's responses. We are confident that, despite the reticence in England to talk about national identity matters (Condor, 2006), such reticence does not imply that the English either do not understand or hold unclear views about the concept. Nor did we find much evidence that the English are baffled by the distinction between English and British (for a full account of the findings, see Chapter 3 and Bechhofer and McCrone, 2010). We use this example here, not to discuss the substantive findings and issues of our research on national identity, but to make the methodological point that handled sensitively, and drawing upon different statistical techniques, survey data can get at aspects of national identity in large samples where other techniques cannot be used.

Survey methodology not only allows us to explore systematically the reasons people give for a category choice, backed up as it is by data from qualitative interviews, but also in repeated and adapted use, allows us to test out our claims about the salience of national identity. Let us give an example of this. Because our work focuses on national, that is, territorial, identity and not on other forms of social identity – social class, gender, age, religion, marital status and so on – it would seem an obvious challenge to our work to say that national identity is far *less* important than we give it credit for. Surely, these other forms of social identity matter more? We have been able to test this out by progressively modifying questions over the range of surveys. We began in the 2001 Scottish Social Attitudes Survey with a question that asked:

Some people say that whether they feel British or Scottish is not as important as other things about them. Other people say their national identity is the key to who they are. If you had to pick just one thing from this list to describe yourself – something that is very important to you when you think of yourself – what would it be?

There followed a list of twenty-three social identities including social class, employment status, gender, age, religion, parental and marital status, ethnicity and so on, including national identity. We then asked people to also give us their second and their third choices. The questions seemed to work well but we will maintain the suspense as to what they showed until the next chapter. Some critics thought that we had perhaps biased the responses by forefronting national identity in our preamble (*'people say their national identity is the key to who they are'*), so when we next asked the question, this time in both the British Social Attitudes Survey, permitting us to look at English respondents, and the Scottish Social Attitudes Survey of 2003, our preamble simply read:

People differ in how they think or describe themselves. If you had to pick just one thing from this list to describe yourself – something that is very important to you when you think of yourself – what would it be?

There followed the same list of twenty-three items as before, and respondents were asked to choose their first, second and third choices. It turned out that dropping the preamble made very little difference. When we tried it for the third time, in 2006 in both England and Scotland, we introduced a more rigorous test: we initially omitted 'national identity' from the list of social identities. Only after the respondent had made their three choices did we ask:

If the list had also included the things on this card [British, English, Scottish, Welsh etc.] would you have chosen one or more of these instead of the ones you did choose?

If the respondent replied yes, then they were asked whether national identity would have been their first, second or third choice. The point of this exercise was to make the test of national identity more stringent each time, so that when we got more or less the same patterns each time we could be pretty sure we had established how important or unimportant national identity really was. By the end of the process, we had far greater confidence in the measure, bearing in mind that the surveys were asked of different people each time, so there was no possibility of respondents feeling a need to be consistent.

We used similar techniques to refine questions relating to people's willingness or otherwise to accept or reject claims relating to national identity. On successive occasions, we have developed the question. In 2003, we asked:[14]

I'd like you to think of someone who was born in Scotland [England] but now lives permanently in England [Scotland] and said they were English [Scottish]. Do you think most people would consider them to be English

[14] This is the version of the question used in England, with the version in Scotland in brackets.

[Scottish]? And do you think you would consider them to be English [Scottish]?

For each question we gave interviewees a five-point response scale ranging from 'definitely would' through to 'definitely would not'. Here the questions asked people to assess just the two markers of birth and residence. We then introduced 'race' and accent into the equation by asking:

And now think of a non-white person living in England [Scotland] who spoke with an English [Scottish] accent and said they were English [Scottish]. Do you think most people would consider them to be English [Scottish]? And do you think you would consider them to be English [Scottish]?

The results confirmed what we expected, namely, that people judged themselves to be significantly more 'liberal' than 'most people', so in future versions, we dropped the first question. In the 2005 survey (Scottish Social Attitudes only), we tried to assess the acceptance of putative claims to Scottish national identity on the basis of birthplace, residence, 'race' and accent. We asked about the respondent's willingness or otherwise to accept a white/non-white person who lived in Scotland while born in England, varying the format according to whether they had a Scottish or English accent. In a set of follow-up questions, we tested the relative strengths of birthplace, accent and ancestry (as measured by birthplace of parents). In the following year, 2006, we were able to ask similar questions in both England (British Social Attitudes) and Scotland (Scottish Social Attitudes), but refined them further. In England, we asked:

I'd like you to think of a white person who you know was born in Scotland, but now lives permanently in England. This person says they are English. Would you consider this person to be English?

The respondent had four possible responses: definitely would, probably would, probably would not, and definitely would not.[15] Respondents (apart from those who 'definitely would') were then asked: *what if they had an English accent?* And again (once more excluding those

[15] This time, we omitted the mid-point in order to force choice.

saying 'definitely would'), *'and what if this person with an English accent also had English parents?'* A similar suite of questions were then asked of a putative non-white person. Thus, by 2006, we had refined and extended the suite of questions to cover birthplace, residence, accent and 'race' in both England and Scotland, as well as being able to focus on respondents who were 'natives', that is, both born and currently living in each country. We will again keep our findings for a later chapter, preferring here to keep the focus on the methodology we used for getting at levels of acceptance and rejection of claims. Although the reader might think the questions are complicated to comprehend, and thus, liable to confuse the respondents, in practice we found they worked well, in large part because of the skill of professional interviewers using CAPI (computer-aided personal interviewing) in face-to-face interviews where it is possible to assess how well the respondent understands what they are being asked. We had one more refinement to add, this time in the British Social Attitudes 2008 and Scottish Social Attitudes 2009 surveys. It might seem an obvious omission in hindsight, but we had no direct measure of the willingness or otherwise of respondents to assent to (or just possibly reject) what we thought of as the 'default' position. Would, as seems obvious, at least in the case of white respondents, the claim be accepted if a person was born in England, lived in England, and had an English accent and parents? This then provided a baseline, with regard to claims both from a white and non-white person. It gave us a benchmark against which to measure 'race' differences, and allowed us to check whether there was a 'race' differential when all other markers were ostensibly equal.

We have tried to show that survey questions, especially if refined over a number of attempts, can be sophisticated measures of the likelihood of acceptance or rejection of identity claims by people with key characteristics such as place of birth, residence, accent and parentage, the markers which we had identified when talking at length with a variety of respondents in our more 'qualitative' work.

We have focused thus far on how we have successively refined questions relating national to other forms of social identity, and assessing how willing people are to accept or reject putative claims from people with identifiable characteristics. Following intensive interviews, we also designed questions focusing on *symbols* of national identity, 'national' as well as British, and whether respondents judge particular *contexts* (sporting, ceremonial, cultural etc.) to be important. More recently, we have designed questions to assess how people perceive similarities and differences with regard to comparator national groups and states. In short, imaginative survey methodology permits rigorous analysis of content and form of national identities, complementing as well as stimulating further enquiry of a more qualitative kind.

We can, of course, use survey research not simply to ask people in general about national identity, but to identify patterns of responses among different groups of people. Surveys can then reveal patterns of which individual respondents themselves may be unaware. For example, we may find that young people are more likely than older people to think that 'national' identity is more important than 'state' identity. As another example, we may find that men are more likely than women to rate national identity highly, but that they do not rate their gender as highly as women do.

We have spent some time discussing the role of survey methodology as one way, albeit not the only one, of getting at issues of national identity. We do so to counter the sometimes expressed view that only more 'naturalistic' and discursive methods are capable of so doing. In this final section of the chapter, we shall strengthen this point by showing how our programme of work on national identity and constitutional change, which was funded by The Leverhulme Trust, was designed. Precisely because national identity is a sociological, political, cultural and psychological phenomenon, it was imperative to bring sociologists, political scientists, social anthropologists and social psychologists together to study it. Different disciplines could bring different perspectives as well as different methodologies to bear. This work has largely been reported in our book *National Identity, Nationalism and Constitutional Change* (Bechhofer and McCrone, 2009b).

We wanted to cover three key levels: the individual, the organisational and the institutional. Accordingly we designed three complementary strands of enquiry: how individuals carry, alter and use national identities; how organisations and work settings act as intermediate milieux with regard to national identity; and finally, how social institutions both shape and in turn are shaped by national identity. The first of these strands used both cross-sectional survey methodology to focus on public opinion at different time-points, and intensive interviews with a smaller number of respondents whom we talked to at some length on up to three separate occasions. Following up our interest in strategic comparisons, we carried out intensive interviews with 'migrants' – people born in Scotland now living in England, and people born in England now living in Scotland – as well as 'nationals', those born and living in the two countries. We were then able to compare the four groups and this proved to be a powerful tool for analysing how individuals made sense of national identity in different contexts. For example, is where you are born a more important determinant of your national identity than where you live? If you are born in Scotland and migrate to England, do you eventually stop thinking of yourself as Scottish, becoming British, even English? And what happens to English incomers to Scotland? In terms of how they do national identity, are they more like those among whom they live, the Scots, or do they remain 'English'? Such qualitative work has proved a valuable complement to the large-scale surveys in Scotland and England, which are good at establishing benchmarks of public opinion, but less good at getting at what individuals mean by particular responses. Because we had a number of survey time-points (2001, 2003, 2006 and 2008/2009), as well as repeat interviews with the intensive interview cohorts, we were able to make much better sense of real-time change, as well as refining our techniques for getting at national identity in the context of events.

In other parts of our programme of research we tried to understand the contexts within which national identity operates. Thus, does national identity matter for people within organisations and institutions? Is national identity important as people go about their daily working lives? We were aware that we could potentially be criticised for wishing our interest in national identity upon those we studied, so our two anthropology colleagues did ethnographic fieldwork, one in a bank and one in a hospital, studying people in their habitats as they went about their business. Both Jonathan Hearn, who studied the bank (see Hearn, 2009), and Nigel Rapport, who studied the hospital (see Rapport, 2008), provided material relating to how people 'do' (or do not do) national identity by looking at circumstances in which it could be regarded as incidental. By using focused ethnographic research in 'mid-level' organisations, we obtained a more 'naturalistic' understanding of national identity, particularly where we wanted to test out whether or not it really 'mattered' to people according to context. Relatedly, other colleagues, Ross Bond and Lindsay Paterson (2005), studied academics to look at the cultural role of universities in 'expressing' national identity, and crucially whether those employed by such institutions saw it that way.

Finally, we chose to include in the programme the institutional level itself, in the form of two cases: the media and economic development agencies, because they are in essence mediating structures where territorial, and in particular national, identity is concerned. Such institutions both shape, and are shaped by, national identities. In some sense, they speak for and to the nation. This is because they are involved in producing and negotiating national identity, and refract these processes back to the population as a whole.

Getting at national identity

National identity may seem to be at once obvious and yet implicit. After all, virtually everyone has one, and yet most are unsure what it means. In the second part of this chapter we have examined the different methodological approaches which enable us to explore its salience and meaning, and have confidence in the findings.

In the earlier part, we have argued that national identity is a claims-making process; it is not a question of having a national identity or not. It does not have any kind of straightforward or singular meaning, and might be better thought of as a site within which argument and debate take place. It is worth repeating here that identity needs to be 'done' over and over again; it is not static, and much depends on how willing other people are to accept a person's claim to be 'one of them'; this is a process of ongoing negotiation. It also has the potential to determine life chances, because if others deem us not to be fit and proper people, then they can deny us our share of society's resources. In other words, there is a complex politics behind claims to national identity. Our task in the next chapter is to try and show that, for the vast majority of people, national identity really matters.

3 | National identity: do people care about it?

Taken as a whole, the overall purpose of this chapter is to ask, when it comes down to it, whether people find national identity meaningful, whether it really matters to them. We will discuss and bring empirical data to bear upon a number of common assumptions:

- That many people do not care, and possibly do not know, about national identity. There is a particularly 'British' (or maybe, as we shall see shortly, 'English') form of this argument, namely, that while it matters to the French, the Germans, the Americans and so on, the people of these islands have a much more relaxed and latitudinarian view of these things.[1]
- That, if it matters at all, 'national' identity is less important to people than other forms of social identity such as their social class, gender or family status.
- That, if anything, people in England and Scotland 'think small', that is, identifying far more with locality, town or region, than with 'nation' or state.
- That national identity categories, as used by sociologists and political scientists, are fairly meaningless, accepted by people if they are asked questions by such social scientists, but not part of their own personal world. We explore that assertion by looking at the accounts of national identity which people give in non-survey settings, as well as by examining the reasons they give for their national identity choices.
- Finally, we show how national identity comes into play in different contexts, be they cultural or political.

---

[1] We are reminded of the doggerel by J. H. Goring in *The Ballad of Lake Laloo and other Rhymes* (1909): 'The Germans live in Germany / The Romans live in Rome / The Turkeys live in Turkey / But the English live at home'.

Let us start with the fairly widespread claim that the Scots, the Irish and the Welsh manifestly care about national identity, but somehow the English do not. As Krishan Kumar observed: 'unlike the French, the English have little tradition of reflection on nationalism and national identity' (2006: 423). Let us look at that claim for a moment. The past-editor of *The Guardian* newspaper Peter Preston asserted that the English have no historical memory:[2] 'History for us is a moribund, inert business. It doesn't bring out boiling passions. We've "moved on" so comprehensively that we don't recall where we came from.' That seems to us quite a claim at all sorts of levels, but for our purposes here it is not its validity but that Preston thought and wrote it, which matters. It is not uncommon to find such a view, but every now and again, one finds by way of contrast that old saw popping up: 'Smile at us, pay us, pass us; but do not quite forget / For we are the people of England, that never have spoken yet.' That plainly carries a latent sense of threat: that poking the English with the equivalent of a sharp stick is not a good idea. G. K. Chesterton wrote that in 1907, and it appeared in his poem 'The Secret People' in 1915. In an interesting essay (*The Guardian*, 9 April 2005), the cultural critic Patrick Wright observed that: 'Chesterton's vision of "secret" England dates from nearly a century ago, but it expresses a way of thinking about identity and change that remains influential to this day.' It sets up the (unspoken) identity of 'us', the silent majority, against any change or threat to the existing order: the identity that does not (yet) speak is inherently conservative.[3] Hence, it has been mobilised by those opposed to constitutional devolution for the non-English nations, those in favour of an English Parliament, and as a comment on the so-called West Lothian Question: by and large, a sentiment issuing from the grumbling Right. Patrick Wright (2005) analysed the poem and set it in context, pointing out that it was a diatribe aimed at Fabian socialists and their campaign against ale, of all things. He observed: 'So this curious Edwardian symbolism grew up, in which beer came to be associated with traditional English freedom, while the joyless over-intellectual Fabian meddlers such as HG Wells and George Bernard Shaw put themselves to bed with warm cocoa.'

[2] Peter Preston in *The Guardian*, Sunday, 19 September 2010. See www.theguardian.com/commentisfree/2010/sep/19/turkey-armenia-genocide-history-passion (accessed 16 July 2014).
[3] We are grateful to our colleague Michael Rosie for this elaboration.

Does any of this really matter or is it a cultural curiosity? It does matter, in the sense that all sorts of implications have been drawn from its assumptions, mainly of a political-constitutional sort. Thus, we find it claimed that, in contrast to Scots, Welsh and Irish, the English all too frequently confuse being English and British, as in the comment by the political campaigner Anthony Barnett, that

The English ... are more often baffled when asked how they relate their Englishness and Britishness to each other. They often fail to understand how the two can be contrasted at all. It seems like one of those puzzles that others can undo but you can't: Englishness and Britishness seem inseparable. They might prefer to be called one thing rather than the other – and today young people increasingly prefer English to British – but, like two sides of a coin, neither term has an independent existence from the other. (Barnett 1997: 292–3)[4]

We will have more to say on this unsubstantiated confusion later in this chapter, but there is a more sophisticated form of explanation of Barnett's puzzle. The late Bernard Crick[5] once wrote an engaging essay called 'An Englishman Considers his Passport' in which he wrote:

For the English to have developed a strident literature of English nationalism, such as arose, often under official patronage, everywhere else in Europe, and in Ireland and Scotland, eventually in Wales, would have been divisive. From political necessity English politicians tried to develop a United Kingdom nationalism and at least explicitly and officially, to identify themselves with it, wholeheartedly. (1989: 29)

Put simply, Crick was arguing that English national identity (or at least its potential expression as 'nationalism') had to be kept underground lest it undermined being 'British'. Kumar makes a related point in a

---

[4] Writing in 2013, Barnett reiterated his view that, when people in England were asked 'Which comes first for you, being English or being British?', 'many simply could not understand the question. They felt equally they were both, at one and the same time, in a way that was inseparable. To ask them to compare their allegiance to Englishness and Britain as if these were distinct identities did not make sense' (Barnett, 2013: 214).

[5] Crick lived in Edinburgh in the latter years of his life, and was an enthusiastic supporter of a Scottish Parliament. His essay on 'The Four Nations: Interrelations' was published posthumously in the journal *Scottish Affairs* in 2010.

different way: 'a strong sense of national identity in France, compared
with a weak one in England, have produced characteristically different
outcomes. The French, many of them at least, think they know who
they are; the English are still searching' (Kumar, 2006: 424). Managing
its own 'missionary nationalism' in these islands, comments Kumar,
has carried with it for the English what appears to be suppression of,
but is actually indifference to, a specific English nationalism. The political philosopher David Miller takes the argument a stage further by
claiming that

> it is even doubtful that there is such a thing as an English national identity
> in the proper sense; there is a cultural identity that finds expression in cer-
> tain distinctive tastes and style of life, but if we look for the hallmarks of
> national identity such as political culture and shared history, we would be
> hard put to find anything that is distinctively English as opposed to British.
> (2000: 137)

The question is whether any of these arguments by Crick, Kumar,
Miller and Barnett are correct. Miller makes a sharp distinction
between cultural identity and national identity. He then defines the
'hallmarks of national identity' in such a way that he feels able to
assert that English and British identity are not 'distinctively' different.
We do not share this view of national identity which is rooted in institutional forms rather than created in everyday interactions.

Barnett does not question the concept of national identity but
asserts, in the absence of empirical evidence, that the English cannot
tell the difference between British and English. If this were correct,
it would have considerable political and cultural implications. Our
research, however, shows that it is not.

The other side of the argument shifts from the English to the
non-English, arguing that somehow they have a stronger and more
straightforward sense of their 'national identity' as opposed to their
citizenship, their state identity. It seems intuitively obvious that a
growing sense of Scottish national identity and a demand for greater
self-government leading to the setting up of a parliament are intimately connected. Intuition is, however, an unreliable tool for scholars.
In this chapter we will look more carefully at the importance, or lack
of it, of national identity to the English and the Scots, the two peoples
who fashioned the British state in 1707, asking not only whether it

matters, but if so, under what circumstances, and to whom. Whether national identity 'matters' or not can be viewed in two ways. First, that it matters to people in the sense that *they* think it does, and this informs their actions. Second, that it matters in the sense that *analysts* say, and maybe show that it does, even though people may not think so, or even be aware of it. This is a distinction familiar to social anthropologists in the form of 'emic' versus 'etic' accounts (Eriksen, 1993: 11).[6] Our focus here is on the former in that we wish to understand how people themselves make sense of who they are.

Perhaps national identity only matters to those who study it, and not especially to those who are studied; that other forms of social identity matter more to people: their social class, their gender, their ethnicity, age, religion and so on. Can national identity compete with these? We have put this feasible hypothesis to the test. In the previous chapter we mentioned that between 2001 and 2006 we steadily refined the questions we used to establish whether national identity was as important to people as other aspects of their identity, and on each occasion obtained similar results. We now present the detailed evidence. The original question in 2001, unlike the later ones, was also asked in Wales in the context of a specific Welsh Election Study, so we shall here first present that three-way comparison (Table 3.1).

Two things are clear. First, that, by and large, respondents in the three countries tend to choose similar social identities such as being parents, having a partner, gender, and employment and social class. Second, Scots are much more likely than the English to choose their national identity, and that being Welsh stands somewhere in between. North of the border, being Scottish ranks second overall to being a parent. It matters to all social groups but there are some variations. Women are more likely than men to choose their gender (by 35 per cent to 22 per cent), but both put being Scottish above that (40 per cent and 41 per cent respectively). Similarly, those who consider themselves working class are considerably more likely than middle-class identifiers to choose their class identity (32 per cent to 15 per cent), but both think being Scottish is more important (47 per cent and

---

[6] Eriksen observes: 'In the anthropological literature, the term *emic* refers to "the native's point of view". It is contrasted with *etic*, which refers to the analyst's concepts, descriptions and analyses. The terms are derived from phonemics and phonetics' (1993: 11 footnote 1).

Table 3.1 *Identity choices in Scotland, England and Wales, 2001*

Percentage who choose identity as first, second or third choice	Scotland	England	Wales
Mother/father	49	48	50
Wife/husband	27	27	27
Woman/man	25	30	22
Scottish/English/Welsh	45	20	33
British	11	27	23
Working class	24	19	23
Working person	29	32	26
Base	1605	2786	1085

Sources: British and Scottish Social Attitudes, 2001; and Welsh Election Study, 2001.

40 per cent respectively).[7] In terms of choices, then, national identity matters, especially in Scotland.

As we explained in Chapter 2, in 2006, to test whether we were leading respondents in a particular direction, we omitted national identity from the initial list of identities, and then in a follow-up question asked respondents whether, if it had been available to them, they would have chosen being English or Scottish or British instead of one of their chosen identities. Almost half of Scots and one-third of the English would have done so. For such Scots, the ratio choosing Scottish rather than British was nine to one; whereas for similar English persons, slightly more chose English than chose British. In other words, we can be pretty certain that Scots rate being Scottish very highly, and more so than the English rate *their* national identity, while at the same time there is much comparability between the two national groups in how they view other forms of social identity. We can refine that statement. Men in Scotland are more likely than women to rate their national identity highly. There is also a clear age gradient in both Scotland and England, in that young people are more likely than older people to do so. In England, for example, 48 per cent of young men choose national identity, but only 28 per cent of older men. In Scotland, the comparable figures are 68 per cent and 40 per cent respectively, suggesting that, while there are different gradients within social categories, the

---

[7] These results are discussed at greater length in Bechhofer and McCrone (2009c: 64–94).

Table 3.2 *How proud are you of being English/Scottish?*

	English (%)	Scottish (%)
Very proud	43	70
Somewhat proud	33	18
Not very proud	8	2
Not at all proud	2	0
Not English/Scottish	13	9
Base	1580	1549

Sources: Scottish and British Social Attitudes Surveys, 2005.

striking feature is still how much higher national identification is in Scotland than in England. National identity does matter in England but it matters much more north of the border.

So we can state confidently that when people are asked to rank various identities, national identity does figure large. However, it could be that *none* of these social identities really 'matter' to them in terms of being important, and that what we are getting is a ranking of the relatively unimportant. We tried a different and more direct tack in two of our surveys (2003 and 2005) by asking respondents how proud they were of being English or (in Scotland) Scottish (on a four-point scale), as well as of being British. Our findings in 2005 are shown in Table 3.2.[8]

The large difference is that the English are far less likely than the Scots to say they are very proud. That said, as many as three-quarters (compared with nearly nine in ten Scots) take *some* pride in being English, and very few take little or no pride. Perhaps the English, reluctant as some would say to enthuse too much over such matters, are just being more restrained. Might it be the case, then, that instead, the English are more enthusiastic about being British?

The comparable figures, again for 2005,[9] show as one might expect, that the roles are reversed with a greater proportion of the English than the Scots saying they are very proud (41 per cent against

---

[8] The distributions for 2003 are not dissimilar, and the conclusions to be drawn are the same.

[9] The distributions for 2003 are once again not dissimilar, and the conclusions to be drawn are the same.

23 per cent). Once again, however, combining 'very' and 'somewhat' proud sharply reduces the differential (79 per cent against 64 per cent). Clearly, English people are more proud of being British than are the Scots, but almost two-thirds of Scots do take some pride in being British. This might come as a surprise, given that so few prioritise being British over being Scottish when asked about their own national identity.[10]

We might expect that there would be something of a trade-off between taking pride in your national identity and being British, but this is not so. Regardless of how they view their own identities, most people north and south of the border take pride in *both* national and state identities. Thus, while 7 out of 10 English people take pride in being English *as well as* being British (2005 data), even among the Scots it is as high as 6 in 10; there is little difference between the two national groups.[11]

So let us sum up what we know thus far. In the first place, national identity figures highly in how most Scots see themselves in identity terms, and a sizeable minority of English people do the same. Second, it seems to be not simply a matter of identity being salient but of pride in those identities, and in that respect we find only a modest Scottish–English difference. Third, as we have just seen, there is no evidence of a trade-off between national and state identity in terms of pride, even for the Scots. Scots and English alike have pride in *both* 'national' *and* 'state' identity, rather than in one or the other. We will have more to say about this in Chapter 8 when we focus on 'being British'.

The sceptic, however, might reply that this is all rather abstract and many people feel little day-to-day attachment to the 'national', still less the 'state' levels of identity. After all, both nation and state are somewhat distant from people's everyday experiences, and it seems a priori credible that more immediate 'local' identities may matter more for most people than the 'national' or 'state' levels. There used to be a childhood game where people nested as many territorial identities as they could think of, so we devised a survey question along these lines (Bechhofer and McCrone, 2008):

---

[10] Using the five-point scale in the Moreno question (see Chapter 2).
[11] Where there is a modest difference, it is that 7 per cent of the English and 16 per cent of the Scots said they were not British.

*Sometimes, for their amusement, children give their address as Home Street,
My area, This town, Localshire, My country, Britain, United Kingdom,
Europe, The World. Thinking about where you live now, which **one** do you
feel is most important to you generally **in your everyday life**?*

*[The street in which you live/The local area or district/The city or town
in which you live/The county or region, for instance, Yorkshire, Lothian or
Glamorgan/The country in which you live, for instance, England, Northern
Ireland, Scotland, Wales/Britain/The United Kingdom/Europe]*

What is striking about the responses is how much locality matters to
both English and Scots. Ten per cent identify with the most immediate
area, 'the street'. One-third of the English and 29 per cent of Scots opt
for the second level, the local area or district. The next level – town or
city – is chosen by 27 per cent of the English and 31 per cent of Scots.
When it comes to 'country' (that is, England or Scotland), only 7 per
cent of the English and 17 per cent of Scots think it is most important.
Again, it is not just a matter of factual importance rather than feeling.
People seem to feel emotionally committed to the localities with which
they identify. As many as 81 per cent of the English and 87 per cent
of Scots feel very or somewhat proud of their local area. On the other
hand, there is a lot of 'pride' invested at *all* levels: city, town or county,
country or state, and once more it is not a matter of feeling proud of
one level and not of the others.

We also asked people: *'If you were abroad and someone who knew
this country asked you "where do you come from?", which one of
the options on this card would you choose?'*. Scots were three times
more inclined to choose 'country' than the English (52 per cent compared with 18 per cent), while similar proportions in the two countries
gave their town or city. The level which matters disproportionately
to the English is county or region (21 per cent compared with 8 per
cent). This is neatly expressed in the saw: 'I'm not English; I'm from
Yorkshire.'

This being so, we would expect that it would be people from
English regions with a strong regional dimension who were more
likely to opt for regional identification rather than national. It is true
that we do indeed find the strongest identification with city/county or
region to occur in three such areas: Yorkshire (49 per cent compared
with English national average of 38 per cent), NE England (48 per
cent) and NW England (43 per cent), but if abroad and asked where

they came from, the proportions saying 'England' in these regions are only marginally less than the mean (18 per cent) at 15 per cent, 14 per cent and 17 per cent respectively.[12] This indicates once more that *both* region and nation are in some sense important to people in England, and might suggest there are different regional ways of being English.

Talking about national identity

Is there any justification for the more radical view that surveys are not good ways at getting at the importance of national identity? People are responding to questions designed and asked by interviewers, and not expressing themselves spontaneously in their own words. Can we be sure that national identity matters without putting words into people's mouths? We mentioned earlier that we derived our survey questions from many intensive interviews, carried out because we see some force in that argument. So let us explore talk about national identity in more 'naturalistic' settings.

The first thing to be said is that for most people, even in Scotland where national identity appears to be more salient, it is largely taken for granted, or, in Billig's term, banal. We have referred to William McIlvanney's 'insurance policy' analogy (we have one, can't lay our hands on it and don't remember the small print). So how do people talk about being Scottish? Here is a comment from one of our Scottish respondents about the 'naturalness' of national identity. Referring to the letter from the interviewer, they said:

But why should I need an envelope through the door to address whether I'm Scottish or not? But then again, it's like breathing, you do it. You don't think about it. It's what you do every day and then somebody says to you 'why are you breathing?' then you've got to stop and think about it. So for me to feel Scottish and then [for] somebody just to come along and say 'why are you

---

[12] In designing the question, we judged it necessary to provide a meaningful context ('if abroad, and asked where you came from'). We accept that saying 'Scotland' may be judged by respondents as more meaningful to the foreign listener than 'Lanarkshire', for example, and that our respondents made such a judgement in their survey reply. It is still striking that the Scots do this more than the English, possibly because we know Scots are more likely to be very proud of that level of territory.

Scottish?' I maybe felt like, it's a superficial thought that I've had before. But even the fact that I've been given a week to think about it, I still think that I'm Scottish because I was born here.

It is that taken-for-granted quality of national identity which is most striking ('it's like breathing, you do it'). Here is another Scot talking about national identity:

It's a sense of identity, isn't it, really? I think that's what things come down to. It's whatever makes you feel comfortable and whatever puts you at ease. If you've got the choice between a scabby, rusty bike and a nice red, shiny bike, you'll choose the red, shiny bike over the other because you can feel proud of it. You can get on it and be the envy of all your friends. Not even the envy, you'll just be the same as all your friends. You have something in which, you have something that reflects, that you're part of, that you're proud to be part of [the] country. I'm so proud of my countrymen, not only those who are living but those that have gone before and the contribution that's been made over the centuries. I'm just so proud to be part of that, albeit that I had no input in it whatsoever. I'm proud, by association.

There are many such examples of Scots talking about national identity in an open, spontaneous and personal way. Here is another, linking self and family:

Oh yes, I feel Scottish. That's something, I suppose, which I've always felt. I've fulfilled something this year, which I had been meaning to do for a long time previously and that was establish more about my family background, a not uncommon pastime. There's many a person, especially as they get older, I think, maybe looks into that. I'd always grown up with a lot of information about my father's family, which I've always been fascinated in and I also had the links with going up, every year, to Morayshire. I've never known as much about my mother's side and there was a particular mystery about her grandmother. Nobody knew the name of her. Who was she? So it's been great fun. Along with my son, involving him in this, we've been into Edinburgh two or three times to dig through records. We've made journeys across ... down to the Borders, up to Morayshire and established lots of very interesting facts. I feel very much part of that. It's given me a sense, also, of not coming from just those two people or those two people, it gives you a sense of how very much everyone is ... there's such a range of people that you are generated from. They were all an incredible mixture.

That is a good example of the personal, the intimate and familial sense of national identity. Our colleagues on the National Identity programme, Susan Condor and Jackie Abell, drew upon the interviews to make the point that 'nationality serves as a vehicle through which to express immediate, authentic, personal experience' (2006: 57). Having carried out the interviews in England which formed part of our project, they contrasted how people in Scotland and England talked about national identity. Referring to 'Jenny' who drew the analogy between national identity and 'the red, shiny bike' in the interview from which we quoted earlier, they commented:

Whereas Jenny, like many respondents in Scotland, presented an image of her sense of self as effectively saturated by, and entirely consonant with, her Scottish national identity, respondents in England almost always represented national identity as something worn lightly, and only partially inhabited. (Condor and Abell, 2006: 69)

We might say, then, that Scots tended to adopt a 'thick' description of their national identity, that is, one elaborated and contextualised, whereas people in England tended to adopt a 'thin' description.[13] Here are some examples from our colleagues' interviews of English people talking about their national identity:[14]

I: Would you ever describe yourself as English? Or does that seem a weird thing to say?

R1:   Er, in some ways, yes, I do think it's a weird thing to say. Erm. Cos what does English mean? You speak English.

I: Mm. Mm.

R1:   Erm. And like, when I, when erm, you fill in forms, new surgery, and this that and the other, and, you know, I always put down, cos English doesn't really seem to mean, mean anything to me, other than you speak English. If they ask, you know, what language do you speak, then, yes, I put down English, but nationality, yeah, I always put write British.

---

[13] The distinction between 'thick' and 'thin' description is usually attributed to the social anthropologist Clifford Geertz (1973). Geertz used examples from the work of philosopher Gilbert Ryle.

[14] We have edited the material slightly to remove pause marks (.) so as to make both sets of interviews more consistent with each other.

And another:

I: What is your nationality?
R2:   I'd say English.
I: English? Why English?
R2:   Well, saying that, British is still fine but there's two things. If you're Scottish, you're Scottish, if you're Welsh you're Welsh, Irish, Irish. But if you ask most Scotsmen they're not British, they're Scottish.
I: Tell me about it.
R2:   Yeah. Definitely. If you ask a Welshman, mostly, well not so much with the Welsh but definitely the Irish, as well, they're definitely Irish, not British. And that, in a way, annoys me a bit because therefore I'm English. I mean like, for instance, football, if it's an English team, I support England. I support other teams as well, but as long as they're not playing England.

Talking about being English seems hedged about by not being something else (Scottish, Welsh, Irish etc.), along with a recognition that there is another category – British – which comes into play. English people do not confuse 'English' and 'British' – *pace* Anthony Barnett – but are well aware that both are part of a nest of identities upon which they can draw. Two other respondents in conversation with each other express the point in an unusual way:

R1:   Well, British is a nationality, English is ...
R2:   Is ...
R1:   Well, beyond us, inherent.
R2:   Yeah. Yes. English is perhaps more genetic, isn't it?

A similarly 'ethnic' point is made more strongly by the following respondent who ventures gingerly into what they see as difficult territory:

I: What would you have to do to be English?
R3:   English heritage, Anglo-Saxon.
I: Says you with the red hair.
R3:   Well yeah, I don't care if that seems a bit flawed, erm, white without sounding racist cos I am not racist, I am not racist at all. If you ask me what my race is I am English, I am white and that's my race I am white English.

This leads the respondent into making a distinction between 'nationality' and 'race':

R3:   It's like, how can I put it? When I've employed different people, like some of my Indian colleagues who I've employed in the past. They're British. They're not English.

I: Why are they not English?

R3:   They are not English. I mean from their race, they're Asian, and Pakistani and whatever. They're not, I can't see them as English. I can see them as British, and part of a British nation and, you know they're proud to be British, great, it's fantastic. It's got to be good for the nation as a whole. But they're not English. English is a race. I don't see English as a nationality.

Surprisingly, perhaps, this same ethnic distinction is also made by a woman of Pakistani origin living in England:

R4:   The reason I wouldn't describe myself as being English is because, to me, English means being white.

I: Right.

R4:   Caucasian, and being, of the, like, tryna think, er, you know, the, er, original er, being er a native of, of England, is what I see as being English. So I would never describe myself as being English, but I would describe myself as being British, because I see that more as meaning that I was born in this country, but if I say English, I always also feel that then, if I say to somebody 'I'm, I'm English', they may say, 'Well, hang on, you're not white, how can you be English', and then they confuse …

I: Have you had that said to you, if you've tried to say 'I'm English'?

R4:   I haven't.

I: No.

R4:   No, I haven't, but I felt, once or twice, I've made the mistake of saying I'm English, and then I suddenly corrected myself, Oh I mean, I don't mean Eng- I mean, British, I was born, then I always say, 'Oh, I was born here, I'm not from here.'[15]

I: Right.

[15] The distinction the respondent makes between 'being born here' and being 'from here' is an intriguing one, separating as it seems to do place of birth and, possibly, ancestry/lineage. We cannot, of course, be sure that this is what she had in mind simply from what she said. There is, however, a view that 'being English' has racial and cultural overtones (see interview with the writer Martin Amis, in *The Guardian*, 18 March 2014).

R4:   So, to me, English means being a native of England, having genera-
      tions of families previous, who have lived in England, rather than
      myself, like I'm just a second generation, actually I'm probably first-
      generation British.

The identity talk in their interviews shows that the English are well
able to distinguish 'England' from 'Britain', and do not draw direct
political-constitutional implications from how they define themselves
(Condor, 2010). Indeed, English people are aware that they are enter-
ing difficult territory, and that they have to tread carefully. We very
much agree with Condor's rejection of the view that the English are
simply apathetic, constituting 'a moral or motivational failure, often
seen to be the product of arrogance, complacency or lethargy' (Condor,
2010: 527).

A manner of speaking or a matter of feeling?

What of the view that people who identify strongly with their 'nation'
do so differently from those who identify with the 'state'? It could
be that 'nationals' – people who say they are only or mainly English
or Scottish – are the people to whom national identity matters more
deeply, whereas those who identify with the state – Brits, as we might
call them – do not invest it with much meaning or emotion. It might be
a manner of speaking rather than a matter of feeling, and it is important not to assume that people mean similar things when they use labels
like Scottish, or English, or British. To pursue this possibility further,
we need to explore what they mean by the labels used in the Moreno
question such as 'English not British', 'more Scottish than British'
and so on.

   For that purpose, we divided respondents into three groups: those
who said they were 'mainly national' (English or Scottish) and not
British and more (English or Scottish) than British; those who gave
equal weight to being (English or Scottish)and British; and those who
were 'mainly British' – more British than (English or Scottish), and
British not (English or Scottish). Why do people opt for the category
they do, and what does it tell us about their attachments to national
identity? As we outlined in the previous chapter, in the 2006 British
and Scottish Attitudes Surveys (see Bechhofer and McCrone, 2010) we
offered respondents a list of reasons for their national identity choice.

Following a preamble which said 'I am going to read out a list of reasons people sometimes give for saying that they see themselves as *English/Scottish not British* or *more English/Scottish than British*',[16] respondents were asked to say whether they agreed or disagreed[17] with a series of statements:

(a)   'I feel uncomfortable about the idea of being British because I want to distance myself from the British Empire and all it stood for.'

(b1)   *For people living in England only*: 'In having to be British, English people too often downplay being English, and I think that's wrong.'

(b2)   *For people living in Scotland only*: 'Being British is too often confused with being English and people don't always realise that there is a difference between Britain and England.'

(c)   'I identify with [English/Scottish] history, traditions and culture.'

(d)   'The values of [English/Scottish] education, [English/Scottish] law and [English/Scottish] community spirit are important to me.'

(e)   'I was born in [England/Scotland] and if you're born in [England/Scotland] you feel [English/Scottish].'

(f)   'I feel more [English/Scottish] now that Scotland has its own Parliament and Wales its National Assembly.'

What makes those who give priority to their national identity over their state identity respond the way they do (Table 3.3)? Above all, such people say they identify with their respective cultures and traditions, and their distinctive institutions. It is not a reaction to Empire, it seems, and not much to do with devolution either. In some respects, Scots are somewhat more assertive of national identity, but it is striking that the English are not far behind. Any suggestion that 'national identity' does not matter to the English, at least those who see themselves as 'nationals', seems to be wide of the mark. Are there, however, any underlying structures and patterns which summarise these variations more economically? Factor analysis is a statistical technique which enables us to identify underlying components of similarities and

---

[16] In each case the interviewer read out the Moreno choice the respondent had already made.

[17] We used a five-point scale from strongly agree to strongly disagree, with a mid-point where they neither agreed or disagreed.

Table 3.3 'National' respondents who agree or strongly agree with the above statements

Percentage who agree or strongly agree	English	Scots	English–Scots
(a) Distance from Empire	11	20	–9
(b1) English downplay English (E)	63	NA	NA
(b2) Britain confused with England (S)	NA	91	NA
(c) Identify with history, tradition, culture	82	91	–9
(d) Identify with education, law, community spirit	86	94	–8
(e) Born in country	88	91	–3
(f) Feel more English/Scottish following devolution	35	29	+6

Note: We have excluded the very small number (around 1.5% of the sample) of respondents who said 'don't know' or did not answer in response to a particular item. As a result the base figures in England vary between 837 and 848, and in Scotland between 944 and 949.

differences.[18] In brief, both English and Scottish 'nationals' choose to prioritise their national over their state identities mainly for cultural and institutional reasons,[19] and not because they are making a 'polit-ical' statement about the break-up of Britain. To be sure, being born in the country matters – especially in Scotland – but that has to be seen as a matter of fact rather than an affective choice, for it is not within anyone's gift to determine their birthplace.

And what of those who try to balance their national and state identities by saying they are equally English or Scottish and British? These number just under half of respondents in England (45 per cent) and just over one-fifth (22 per cent) in Scotland. After a similar pre-amble referring to the reasons for thinking themselves equally English/ Scottish and British, we asked them to agree or disagree with the fol-lowing questions:

[18] Details as to method and results are given in Bechhofer and McCrone (2010).
[19] As measured by items (c) identifying with history, tradition and culture, and (d) identifying with education, law and community spirit.

(a1) *For people living in England only*: 'You can be equally proud of being British and of being English; it's not a matter of choosing between them.'

(a2) *For people living in Scotland only*: 'You can be equally proud of being British and of being Scottish; it's not a matter of choosing between them.'

(b1) *For people living in England only*: 'It is important to me to recognise that England is an equal partner with the other countries in the United Kingdom.'

(b2) *For people living in Scotland only*: 'It is important to me to recognise that Scotland is an equal partner with the other countries in the United Kingdom.'

(c1) *For people living in England only*: 'Sometimes it is more appropriate to say you are British and sometimes it is more appropriate to say you are English.'

(c2) *For people living in Scotland only*: 'Sometimes it is more appropriate to say you are British and sometimes it is more appropriate to say you are Scottish.'

(d1) *For people living in England only*: 'Britain has an important history in which England played a significant part.'

(d2) *For people living in Scotland only*: 'Britain has an important history in which Scotland played a significant part.'

(e1) *For people living in England only*: 'I feel English as well as British now that Scotland has its own Parliament and Wales its National Assembly.'

(e2) *For people living in Scotland only*: 'I feel Scottish as well as British now that Scotland has its own Parliament and Wales its National Assembly.'

(f)   *(asked in England only)* 'There is **no** real difference between English and British.'

The reasons given by the English and the Scots for choosing equal identities are again very similar (Table 3.4). Using factor analysis once more, we find that the first component in both countries attributes the choice to the view that there is no need to choose between them, together with the response that each country is an equal partner in the Union. These respondents have a sense that different identities have their place in different contexts. There seems also to be a more 'instrumental' association between national and state identities. For

Table 3.4 *Respondents who are equally British and English/Scottish agreeing with the following statements*

Percentage who agree or strongly agree that:	English	Scots	English–Scots
(a) Not necessary to choose – one can be equally proud of both	93	97	–4
(b) Equal partner with other countries in UK	86	93	–7
(c) Sometimes more appropriate to be British or English/Scottish	56	67	–11
(d) England/Scotland played significant part in British history	86	93	–7
(e) Feel English/Scottish as well as British following devolution	49	52	–3
(f) No real difference between English and British	67	NA	NA

Note: We have again excluded the very small number (around 1.5% of the sample) of respondents who said 'don't know' or did not answer in response to a particular item. As a result the base figures in England vary between 1042 and 1059, and in Scotland between 274 and 275.

the Scottish group, the sense of British history is associated with the concept of Union as partnership, and hence their 'Scottishness' has an 'affective' association with being British. Devolution figures more prominently for those who describe themselves as equally English or Scottish and British than for the other identity groups. This suggests that they interpret devolution not as a step on the road to a Scottish state but as conferring co-equal status and partnership on the two countries, making it sensible to hold national and state identities equally.

Finally, there are the small proportions of 'Brits' – only 14 per cent in England and a mere 3 per cent in Scotland – who give priority to being British by saying they are 'more British than (English or Scottish)' or 'British not (English or Scottish)'. Once more, we asked them to agree or disagree with six questions, using the five-point scale previously described:

(a) 'Being British is important to me because all parts of the United Kingdom are included.'

Table 3.5 *'British' respondents agreeing with the following statements*

Percentage who agree or strongly agree that:	English	Scots
(a) All parts of the United Kingdom are included	86	96
(b) Should celebrate past achievements of GB	72	82
(c) Includes all ethnic minorities and people of different cultures	76	73
(d) Identify with monarchy, British traditions and ceremonies	64	73
(e) British Empire was an important part of our history	55	70
(f) Feel more British following devolution	20	6

Note: The base numbers in Scotland are very small (less than 100) and the percentages given are only indicative. We have once more excluded the very small number (around 1.5% of the sample) of respondents who said 'don't know' or did not answer in response to a particular item. As a result the base figures in England vary between 298 and 304, and in Scotland between 44 and 46.

(b) 'I think we should celebrate the past achievements of Great Britain.'
(c) 'Being British brings us together because it includes all ethnic minorities and people of different cultures.'
(d) 'I identify with things like the monarchy, British traditions and ceremonies.'
(e) 'Being British matters to me because the British Empire was an important part of our history.'
(f) 'I feel more British now that Scotland has its own Parliament and Wales has its National Assembly.'

For the 'English' Brits, we see that devolution has far less impact than it had for those who saw themselves as equally English and British (Table 3.5), and where it does, it implies a desire to cement together the United Kingdom. What factor analysis shows for this group is that there are different ways of being 'British'. Those with a considerable attachment to 'empire loyalist' views ((d) and (e) in particular) contrast sharply with those whose Britishness takes a more 'inclusive' dimension (responses (a) and (c)), that is, those of a more 'liberal' disposition. When people in England say they are 'British', their reasons for giving this view can be very, even diametrically, different.

Identity in context

In this chapter we have discussed how important (or otherwise) national identity is to people, and we have tried to show what lies behind their responses. Our final task in this chapter is to see whether there are circumstances and events which make us more aware of our national identity. Here, we are responding to an obvious retort to those of us who study national identity: it's unstable; it varies all the time in different situations. It should be noted here that this is a very different position from our emphasis on identity in context. These critics are implying that national identity varies almost randomly in different contexts as opposed to our more structured view. In the British Social Attitudes Survey of 2008 and the Scottish Social Attitudes Survey of 2009, we explored this by asking people in England and Scotland to tell us whether they would feel more or less 'national' in different kinds of situation. We asked:

Some people feel more English/Scottish in some situations and less English/Scottish in others. Where would you put yourself on this scale in the following situations:[20]

When the national team are playing in a sporting event?
When you are in Scotland/England? (i.e. the "other" country)?
When you hear or read about the Scottish Parliament or the National Assembly of Wales?
When you see images of or visit the English/Scottish countryside?
When you hear, read or look at famous English/Scottish music, poetry or paintings?

Respondents were shown a five-point scale: 'feel a lot more Scottish/ English; feel a little more Scottish/English; feel about the same; feel a little less Scottish/English; feel a lot less Scottish/English'. The first thing to say about 'nationals', whether in England or Scotland, is that, remarkably, the rankings are identical, although there are clear differences between the countries in terms of intensity (Table 3.6).

---

[20] This question was asked of people who described themselves as English/ Scottish not British, more English/Scottish than British, equally English/ Scottish and British, or more British than English/Scottish. It was not asked of those who said they were British not English/Scottish.

Table 3.6 *Contexts in which people felt more national*

Percentage who say a lot more or a little more national when:	English	Scots	English–Scots
National sports team playing	65	71	–6
Seeing images of/visiting countryside	54	66	–12
In other country (England/Scotland)	52	65	–13
Experiencing music, poetry, paintings	44	61	–17
Scottish Parliament or National Assembly for Wales	40	30	+10
Base	2138	1256	

Sources: British Social Attitudes Survey, 2008; Scottish Social Attitudes Survey, 2009.

It is interesting, and perhaps surprising, that when considering the context of devolution for Scotland and Wales, it is the English, and not the Scots, who are stimulated to feel more 'national'. We should not make too much of this though, because it is very much a minority pursuit. Less than one-third of Scots say they feel more Scottish when contemplating devolution, which fits with what we have said elsewhere about devolution having very little direct impact on national identity north of the border (Bechhofer and McCrone, 2010). Scots are also more likely than the English to say they feel a 'lot more' Scottish when the national sports team is playing, and on the occasions when they visit England.

Can we say what kind of person feels 'more national' in these contexts? It could, a priori, be a function of age, sex (men and football?), education or social class as well as national identity. Further statistical analysis strengthens the general argument we are developing that the most important influence on feeling more national in the contexts we examined is national identity.[21] Whether in Scotland or in England, those who already have a strong sense of national identity are most likely to say they feel more national in these contexts.

[21] The technique we used is binary logistic regression. National identity is by far the most important variable. Being older has a strong effect in both nations with regard to listening to music, reading poetry or looking at paintings, and devolution and in England with regard to seeing images of or visiting the English countryside.

Table 3.7 *Contexts in which people felt more British*

Percentage who say a lot more or a little more British on:	English	Scots	English–Scots
Sporting occasions	68	60	+8
National anthem	60	32	+28
When abroad	58	38	+20
Ceremonial occasions	49	30	+19
Base	1864	797	

Sources: British Social Attitudes Survey, 2008; Scottish Social Attitudes Survey, 2009.

We also put a similar set of scenarios to those who felt British in some way, excluding only those who said they felt only 'national', that is (English or Scottish) and not British. The questions were as follows:

Some people feel more British in some situations and less British in others. Where would you put yourself on this scale in the following situations:

When you hear the national anthem ("God Save the Queen")?
On sporting occasions like the Olympics when a British team is competing?
When you go abroad?
On ceremonial occasions involving the monarchy, such as royal weddings or the state opening of parliament?

Our findings, using a similar format as for the 'national' situations, are shown in Table 3.7.

The rankings are again very similar, but this time there are striking differentials between the English and the Scots. 'Sporting occasions' make a majority of Scots (60 per cent) and the English (68 per cent) feel 'more British'. However, most of the Scots feel only a 'little more' (38 per cent) rather than 'a lot' (22 per cent). The English, on the other hand, tend to reverse those proportions (28 per cent to 40 per cent). The other three contexts all affect the English more than the Scots and, moreover, in each of the three the English are more likely compared with the Scots to feel a 'lot more' British as opposed to a 'little more British'.

If we once again model the data statistically, in the case of the English, it is age which has the greatest effect: the older you are, the more likely you are to say you feel more British in these various contexts, above all with regard to the 'national' anthem and British ceremonial occasions. On the other hand, in contrast to the previous model, national identity, feeling British has *little* discernible effect, except weakly on sporting occasions, and ceremonial occasions.

What of the Scottish Brits? There is a problem because relatively so few say they are 'mainly British' in the Moreno format, and the findings have to be treated with caution, but we find strong effects[22] relating to being older, national identity (being British), and being female. Sex and national identity have strong effects in all our 'contexts' apart from 'British sporting occasions'. What this suggests is that, in Scotland, for the group who feel British it is positively associated with hearing the British national anthem, watching British ceremonial and being a Brit abroad, all seemingly reinforcing one's sense of being British. In England, on the other hand, only age has anything like this effect on reinforcing Britishness, but that lack of association possibly reflects the fact that 'British' in England has two very different aspects as we have already seen – a 'conservative' one associated with tradition and history, and a 'liberal' one, associated with multiculturalism.

What has this chapter added to our understanding of national identity? National identity, it is clear, is neither a badge to be displayed nor simply a flag to be routinely waved. In an important way, it is deeply personal and internal. It matters to people because they are bound up in it. The Norwegian social anthropologist Thomas Eriksen observed: '"nation" is the metaphysical space in which people locate their personal histories, and thereby their identities' (in Cohen, 2000: 152), so that personal identity becomes synonymous with national identity. It is not the case that people simply buy into what is offered to them *de haut en bas*; they actively construct who they are from what is available to them. This is why national identity is not rigidly fixed but varies systematically according to context and the markers such as ancestry, residence and commitment which people can deploy in the situation in which they find themselves.

---

[22] Using binary regression models as before.

In asking whether or not national identity matters to people, the evidence we have presented in this chapter suggests that it does and that they care about it. It is not an empty category dreamed up by survey analysts and government bureaucrats to which they respond without giving it much thought. This is so, we conclude, because national identity is highly personal, that, in the words of the American novelist James Sallis, 'like nations, individuals come to be ruled by their self-narratives, narratives that accrue from failure as much as from success, and that harden over time into images the individual believes unassailable' (Sallis, 2008: 59).[23]

So national identity matters because it is insinuated into people's own sense of themselves, to the point that, in Bruce Kapferer's words, 'the person conceives of self as also the nation' (1988: 161). Condor and Abell employ the useful metaphor of 'lamination' which wraps around personal and collective identity, the one reflecting the other (2006: 58). Anthony Cohen has observed that the concept of the nation is something standing outside ourselves but is:

(s)omething which simply does not require to be well defined, first, because people presume that they know what they are talking about when they refer to it; and second, because the lack of definition allows them scope for interpretive manoeuvre in formulating or inventing or imagining the nation in terms of their selves for the purposes of personal identity. (Cohen, 2000: 166)

This process of carrying out 'interpretive manoeuvres' is often most apparent in situations in which national identity is debatable and open to choices and challenges. Accordingly, it is to one such context that we now turn.

[23] We are grateful to Sean Damer for this reference.

4 | Debatable lands: national identities on the border

They talk about Scotland and England indeed;
there's Scotland and England and Berwick on Tweed.

Pay a visit to the tourist office in the small town of Berwick-upon-Tweed on the Scottish–English border, and you encounter some curious identity tales. Berwick 'is at war with Russia'; it has its own language: gadgie (a man); manashee (woman); jougal (dog);[1] it changed hands between Scotland and England some fourteen times; there is also the ditty about Scotland, England and Berwick, and many more. Many small towns have developed their distinctive curiosities to bring in the tourists, so why should Berwick be any different? And why in sociological terms is this interesting? More to the point, what does it tell us about national identities?

Berwick, a town of 11,500 people, sits in the borough of Berwick-upon-Tweed, the most northerly in England, with a population of 26,000. Berwick is further away from London than anywhere else on the mainland, even further from London than Penzance in Cornwall by some fifty miles. In jurisdictional terms Berwick's regional affiliations are to Northumberland, the most northerly county in England, and it lies some 50 miles from Newcastle, Northumberland's largest city which has a population of around 190,000. Berwick is at the far north-easterly end of the Scottish–English border which, far from running from west to east, slopes up roughly north-east from its

---

This chapter is based largely on previously unpublished data and we wish to acknowledge the great contribution made by Richard Kiely, the research officer working on our identity projects, in terms of data gathering, analysis and some drafting. See Kiely et al. (2000) for an earlier account of this research.

[1] These terms seem to have been adopted from 'Romany' or travelling people who historically travelled up and down the east coast of Britain in pursuit of the horse trade among others.

south-west end, and is further north in terms of latitude than much of
southern Scotland.

The current borough of Berwick-upon-Tweed, its English nomenclature, straddles the river Tweed. 'Berwick' is the oldest part of
Berwick-upon-Tweed and lies within the Elizabethan town walls north
of the river. South of the Tweed lie Tweedmouth and Spittal. These latter
two parts of the town were originally separate villages but joined with
'Berwick' to make up Berwick-upon-Tweed in the nineteenth century.
Unlike 'Berwick', these villages were never part of Scotland. Berwick
has been within the jurisdiction of England since 1482. It has changed
hands between England and Scotland fourteen times in its history; was
one of four 'royal' burghs of Scotland with, until 1482, special trading
rights conferred by the Scottish Crown; and was subjected to slaughter
at the outbreak of the Wars of Scottish Independence in 1296 when
between 8000 and 15,000 of its inhabitants, virtually the whole population, were killed by the invading English army of 'Longshanks', King
Edward I. The town and its castle were occupied and reoccupied by
Scots and English successively, long after the Scottish Declaration of
Arbroath in 1320 had brought the 'hot' war between the two countries
to an end. Retaken by the English in 1482, the walls were strengthened
under Elizabeth I of England in the 1550s in order, as the (English)
tourist board currently puts it 'to keep out the marauding Scots who
regularly laid claim to the town'; reminding us that history gets written by the victors, even after 500 years.

Being the subject of war helps to keep reputations alive, as does the
curious tale that Berwick is still at war with Russia. The claim is based
upon a widely held belief that the town had been mentioned separately
in the declaration of hostilities at the outbreak of the Crimean War
in 1854 but omitted from the Peace Treaty. The Borough Secretary
lays to rest this myth, one of many in Berwick, in a booklet produced
by the local council on the issue titled 'Berwick-Upon-Tweed's war
with Russia'. It concludes that 'this is yet another of the many romantic fictions which have surrounded the town throughout its long
and chequered history' and states that the UK Foreign Office, when
consulted about the issue in 1965, confirmed that their library had
investigated the matter in the 1930s and found nothing to support the
story. Nonetheless, this 'romantic fiction' is seen by many people in
Berwick-upon-Tweed to sum up neatly the uniqueness of their town
and its identity. In many ways it is a silly tale, but silly tales can reflect

sociological if not historical reality and it helps to mark out the curiosity value of this most northerly English town, as well as being a boon to the passing tourist trade. So is the notion that there is still in Berwick the jurisdictional border between England and Scotland, thus allowing the enterprising tourist to be photographed standing astride the border, a foot in each country. Sadly this is not so because the border is no longer marked by the wide river Tweed at Berwick, but is three miles north on a windy hillock at Lamberton. The intervening space was designated as a *cordon sanitaire* by the English state in the fifteenth century, and given over to the 'Guild of Freemen' of the borough with 'privileges, customs and responsibilities', and so it has remained down to the present day.

Borderlands are interesting precisely because they are on the periphery. The historian Stefan Berger has commented that 'within national histories, borderlands play an important role, as it is at the border that the nation defines itself most rigorously' (2009: 497). The example he gives is Karelia, the region between Finland and Russia, and famously the subject of the suite by Jan Sibelius which became a potent nationalist emblem for the Finns. Borders and frontiers continue to be significant. This is undoubtedly the case in Scotland. Even if, happily, they do not loom quite as large as in, say, the former Yugoslavia, the implications of a state border between England and Scotland in the event of independence were hotly debated during the referendum campaign.[2] On the one hand, borders can grow in political importance as governments seek to restrict the flows of goods and people across them. At the same time, borders between states are reduced in significance by common agreement for political and economic purposes. These opposing tendencies lead the European Union to seek to strengthen its external perimeter (its 'thick' boundary), while encouraging cross-border flows between states – its 'thin' boundaries. Most research on borders focuses on boundaries between states. We know far less about boundaries between 'stateless' nations such as Scotland and England, even where they once had considerable political and cultural significance when these countries were formally independent. Partly this is because interest centres on those which are currently the subject of serious dispute and around which, and over which, conflict is generated.

---

[2] See, for example, www.scotsman.com/news/politics/top-stories/ed-miliband-warns-of-independence-border-posts-1-3459730 (accessed 16 July 2014).

Such conflict-prone borderlands frequently have a key role to play in making and sustaining national identities. There are many examples of contested lands: Alsace-Lorraine between France and Germany post-1870 German unification; the Sudetenland between Czechs and Germans in 1939; the *krajina* (literally, 'borderlands') between Serbia, Croatia and Bosnia in the implosion of Yugoslavia in the 1990s; and, closer to home, the lands between the Irish Republic and Northern Ireland before the peace accord at the end of the twentieth century. One might of course argue that, because Scotland and England are part of the same state (since the Treaty of Union in 1707), this is not such a meaningful border in political-legal terms any more. But it is; not simply because since 1999 there is a (domestic) Scottish Parliament, and since 2007 a nationalist government in Scotland, but because there are continuing and meaningful jurisdictional differences including legal ones; on one side of the border Scots law runs, and English law runs on the other. To be sure, no-one, thankfully, gets killed these days for belonging to the 'wrong' nationality, but ambiguous Berwick does provide the opportunity to understand national identity.

Social interaction around the border sheds light on what it means to be 'Scottish' or 'English'. We have a unique opportunity to comment on processes of identity formation, maintenance and transformation in relation to a border, precisely because it is not the subject of conflict, and thereby remains a fluid and problematic boundary. In the last decade of the twentieth century we grew more accustomed to hearing about towns on the border. In many instances they become problematic and dangerous places, places of contested and ambiguous identities in which the inhabitants exaggerate their ethnicity, and become more ethnically 'pure' precisely because of their frontier status. We can think of the former Yugoslavia – Vukovar and Knin – or even the ethnic and national tensions of the border towns of North and South Ireland, Newry and Londonderry/Derry.

We are not suggesting that Berwick is remotely likely to become a crucible of conflict, but it does provide an opportunity to examine how identity politics operate. The relative lack of direct conflict is important because we have learned from Michael Billig that national identity is 'banal'; it is basic, important, but also taken for granted. Thus, the Scottish–English border is banal. It could be that actually the border matters very little, even at all. After all, those who live there have to make a living as best they can, and this involves precisely those kinds of social and economic transactions that the rest of us make, without worrying about whether those with whom we have dealings are not 'like us', not 'our people'. As we pointed out earlier, the European Union's 'thin' boundaries are predicated precisely on such a view. Put simply, we might expect that people living in Berwick and its nearby communities have more in common than they have differences. Indeed, we might even find that they stress these commonalities at the expense of differences simply because to do otherwise would make life more difficult. This might involve finding and sharing identities, whether they are local, regional or even national. As far as these are concerned, saying you are a Berwicker (local), a borderer (regional) or even, if more problematical when looking north, British are some of the possibilities. On what grounds might people claim to be English or Scottish, on the other hand? 'English' is perhaps the most obvious because after all, Berwick is geographically situated in England, is governed by English laws and institutions, and the border merry-go-round stopped moving in 1482, certainly sufficient time for people to get used to being English. Are there any grounds at all, then, for those in Berwick claiming to be Scottish? Migrating from north of the border might be one such, or having Scottish-born parent(s). More contentiously, it might take the form of claiming that Berwick had been an important burgh of Scotland, and that it had lost that identity by forcible conquest and not persuasion. We are in the realms here of *Scotia Irredenta (*literally, 'Scotland unredeemed'*)*, especially as the old town of Berwick remains on the 'Scottish' side of the Tweed, the historic border, with the settlements of Tweedmouth and Spittal on the south, or 'English', bank.

The point here is not to search out the 'correct' answer to the national identity question. There is no such thing; rather, it depends on the claims one can muster and how well (or badly) these can be made, and crucially, what significant others judge these claims to be. Let us start with the 'legal' position, that Berwickers are English because the town is in England. Here is one such claim:

I: Can I ask you what your nationality is?
R1:   See, I would say English. If I was writing it down I would say British, but if anybody asked me I would say English.
I: Can I ask you why that is?
R1:   You see, I've spent years trying to tell people I'm English, like when I was in the RAF, there was all this basic training, I was the only one sort of from north of Newcastle so I was always called Jock. So

I spent like a year consciously persuading people I was English. So I have always made a point in making sure that people knew I was English.

This is an interesting account because the respondent alludes to the English/British distinction (the latter reserved for 'writing it down'), and how on being 'taken for' a Scot ('Jock') by colleagues in the RAF, he has gone out of his way to assert being English.

The contextual nature of identity is expressed in the following account:

R2: Personally, I mean, if anybody asks me, I'm English because Berwick is part of England, this is where I was born, and to me it would be as ridiculous to say I was Scottish as somebody from Eyemouth [8 miles to the north and in Scotland] who said they was English. Because it isn't an actual fact.

I: Would you describe yourself as a Berwicker?

R2: It would depend who I was talking to. If you're away further south or something I wouldn't sort of describe myself as a Berwicker.

I: How would you describe yourself?

R2: Well, English, basically.

I: If you were talking to someone in the town, for example, would you describe yourself as a Berwicker?

R2: Yea, if you was talking to somebody, say fairly local, not necessarily in Berwick, but say Eyemouth, Wooler [16 miles south] or that sort of area.

Here we find being English asserted as a matter of 'actual fact' ('because Berwick is part of England'), just as residents of Eyemouth over the border are 'Scottish' whether they like it or not. Further, the context matters ('It would depend who I was talking to'), so south of Berwick the identity claim is to be English but a different identity claim (being a 'Berwicker') would be made locally in towns such as Eyemouth and Wooler. It is interesting to note that they are on different sides of the border but the speaker does not see this as relevant because he believes they are close enough to Berwick for the claim to be meaningful.

Being 'taken for' a particular national identity is something people from Berwick have to confront. Here is an exchange between husband and wife which makes the point well and highlights that living near the border can make people feel strongly about their identity.

R1:   Well we're definitely English but you stand up for your identity as a person from Berwick, which is quite strong.

I: Do you feel more strongly a sense of coming from Berwick than being English?

R1:   No; definitely English.

R2:   If he [R1] goes, even to Newcastle, he is classed as being Scottish and he hates that.

I: Do you know why you feel such a strong reaction to people …

R1:   It's because I'm English and because I live so near the border. When you go into Scotland they can pick up the Scottish accent but they know you are not Scottish. I mean it's blatantly obvious. But when you go into England you just sound Scottish, you are branded as being Scottish. Even though you're not. And I'm as patriotic as St George!

And later in the same interview:

R1:   People have said that the north side (of the town) is Scotland. It's not. The border is another 3 mile north, but there are people say it was the border and it should still be.

I: That is not something you would agree with?

R1:   I would not agree with, no. 'Cos I was, I was born in Castlehills [maternity home] which is on the north bank of the Tweed, ok? And I was brought up and I lived 'til I was 23 on the north side of that and there isn't a fibre in me, of my body, that is Scottish. So if anybody said, 'oh that should be Scotland' then I would sternly disagree.

I: Do you think, living where you do close to the border makes you more aware of your nationality?

R1:   Definitely; without doubt.

Here we have further clues about national identity in Berwick. The residents have to respond to being 'taken for' Scottish because of perceived accent, or being born and living on the north bank of the Tweed, in 'old' Berwick. Berwickers are keenly aware of how others see them, however mistakenly as far as they are concerned.

We's fae Bereek!

So what sort of national identity claims do people in Berwick make, and why? One clue lies in our opening ditty: Scotland, and England *and* Berwick on Tweed. Here are some examples:

I: If I was to ask your nationality?
R: Well, Berwicker, sort of, really in between, I can't say. I'm a Berwicker, never sort of 'I'm English, I'm Scots'. Just being in the toon all my life, like, you're neither Scotch nor English, so you're a Berwicker.

And another, talking about locals:

They're Berwickers, they're nothing else but Berwickers, they don't associate themselves either with England or Scotland. I've heard that said at work, on quite a few occasions people have said, 'Oh I'm not English or Scottish, I'm from Berwick.'

Moreover, some incomers claimed that locals from Berwick-upon-Tweed used this strategy of localism to try and avoid or to overcome the problems of 'being national' in this unusual identity context.

There is still quite a strong tendency not to get caught up in the national-ist thing. It's probably outsiders coming in that tend to try and put them in one camp or the other. And I think they do regard themselves as being Berwickers and they are neither Scottish nor English. They are slap bang on the border and I suppose they can't afford to be either really.

Stressing localism can then be a way of sidestepping the problem of being national – English or Scottish. Here is a dialogue between two people talking to the interviewer, one (R1) describing herself as a 'true Berwicker', and her husband (R2) who is not originally from the town.

R1 (Mrs):   It's part of the uniqueness of Berwick, not Scottish not English, but Berwick.
I: People in Berwick, would they generally say they were English?
R1 (Mrs):   I don't think they'd say they were English.
R2 (Mr):   Northumbrian?
R1 (Mrs):   Northumbrian maybe.
R2 (Mr):   But it's Berwick!
I: So when you say, they wouldn't say English, I mean, they wouldn't unless pushed, or they just wouldn't …
R2 (Mr):   If you're pushed, I think, you would find people would say they are Northumbrian, but otherwise, I think they would say Berwick.
R1 (Mrs):   England to me is the south, and I can't bear it.

What this exchange reveals is again the importance of the context in which claims to be a Berwicker are made. Above all, it is set within a Scottish–English frame, while it also introduces a regional identity – Northumbrian. One also finds that family and ancestry figure in claims, as in this exchange:

R3:   My mother, all her side were from the [Scottish] Borders, and she gets quite annoyed if I say I'm English: 'You're not – you're half and half!'
I: So Berwickers …
R3:   … It's to be not-English, not-Scottish; they want to be their own. And I mean, there is so many families, I think, like mine: one or other is English, or Scottish.
I: So you say 'Berwicker' is to sidestep …
R3:   Yes, I think so. I think that's where it comes from.

This propensity to claim local identity in preference to national ones can be seen more systematically from survey data collected after the qualitative interviews. Thus, roughly half (47 per cent) of the 115 people we interviewed in Berwick described themselves as 'Berwickers', and of these by far the largest proportion – 82 per cent – said they *always* or *usually* used the term 'Berwicker' (as many as 61 per cent said they always used it). Feeling you have something in common with Northumbrians on the one hand (68 per cent) and with [Scottish] borderers on the other (67 per cent) reflects local propinquities and interests, and there is such reluctance to look further afield that only around 3 in 10 see commonalities with national, that is, English and Scottish identities. It is clear that many Berwickers are forefronting their local identity. We get similar distributions if we ask about the '*importance*' of these territorial identities. Thus, of those describing themselves as 'Berwickers', almost two-thirds (64 per cent) say it is important to them (and 34 per cent say it is 'very important'). Those describing themselves as 'Scots' are more likely than 'the English' to say it is important to them (respectively, 58 per cent and 50 per cent), with almost twice as many of the former saying it is 'very important' than the latter.

It seems, then, that claiming to be a Berwicker is an important and frequently used form of social identity; and that it is used to prioritise one's identity rather than totally deny 'national' identities, which are handled by recognising that 'Berwick' has both English and Scottish connections in identity terms. More people are 'proud' of being Berwickers than any other identity, but being Scottish and English also figure, and in minor key, local regional identities – being Northumbrian or borderers.

So what makes a Berwicker and what markers are used to justify the claim? Family ties are important sources of claims, and two-thirds identify having parents born in Berwick, followed in significance by having parents living there or living there oneself. Having a local dialect is also something which figures quite prominently in people's accounts of being 'Berwickers'.

Making a Scottish claim is by no means unknown in Berwick. On what basis do people do that, given that jurisdictionally Berwick is in England? Here are some examples:

R1:   I think it is so stupid when you look at the map, and then the border goes away [from the river]; where it should just follow the river – to me it's the most natural thing. And after all Berwickshire [the county] is in Scotland isn't it, so why not Berwick? I think we were captured by the English!

R2:   My father, well he originally come to Berwick through the army, the King's Own Scottish Borderers (KOSBs). That is how he ended up here, like. But I've always had well strong feelings for Scotland because my father come from there, like. And I was in the King's Own myself. It was a *Scottish* regiment. You see, there you are, it's called [King's Own] *Scottish* Borderers.

R3:   I've been inclined to think I was Scottish.

I: Even though you were born in Berwick?

R3:   Yes, because Berwick was in Scotland originally. So, where do you start a book? You start at the beginning, not in the middle, and everything points towards it being Scottish. The times when the English took over, over the years all they've done was massacre the town people ... It was a bustling port, Berwick was bigger than Edinburgh, you know. It was looked upon as the capital of Scotland, so historically, I think we should belong to Scotland rather than England.

These comments illustrate the grounds on which some inhabitants of Berwick claim to be Scottish: that the 'natural' border is the river Tweed, and if that were still so, (old) Berwick would be in Scotland (by implication, leaving Tweedmouth and Spittal on the 'English' side of the border); that the local regiment, with headquarters in Berwick, was a Scottish one; that many family origins are Scottish; that Berwick was historically a Scottish town forcibly taken by the English in war.

Let us emphasise, because it is such a crucial point, that there is no 'correct' answer to these national identity conundrums. Sociologically speaking, there is a multiplicity of criteria which allow claims to be made. For example, in claiming to be 'English', residents can cite jurisdiction going back over 500 years; being educated under an English school curriculum; governed by English civil institutions; having English-born relatives and so on. On the other hand, being 'Scottish' can be based on 'history', conquest, location – being on the 'right' side of the Tweed, family origins and so on. Many people assert their local identity – being Berwickers – and are certainly aware, sometimes uncomfortably aware of being 'taken for' one or other national identity (in terms of accent and language, for example). Here is an example: 'If you go to Newcastle they think you're Scotch; if you go to Edinburgh they think you're English. And I think Berwick is pretty unique, and just ... you're a Berwicker!' It is clear that the way to resolve the competing claims is simply to assert local identity; in the words of another respondent, making use of local dialect ... 'We's fae Bereek!'

There are many places, towns, villages and cities, which assert the importance of locality. We saw in the previous chapter that most people in England and Scotland – upwards of 80 per cent – think that their town, district or street matter most to them. So is Berwick any different? Why should we see claims to localism in Berwick as in some ways unusual and special, and a reflection of national identity conundrums?[3] Perhaps we are making too much of Berwick's unique and problematic status and it is just the concern of a few people in the town and few else.

How others see them

In this section, we will explore how others see the Berwickers, because if we are correct, we would expect those significant others to confirm

[3] A mile or two north of Berwick, but still within the English side of the border, there is a farm called Conundrum, close to the battle site of Halidon Hill (1333) at which the Scottish army seeking to relieve the siege of Berwick was defeated by the English army defending the town. We have been unable to find the reason, if any, behind the name, but it seems to us an appropriate descriptor of the status of Berwick in national identity terms.

and reinforce the problematic identity status of the town. If, on the other hand, the residents of neighbouring towns and villages also convey their own intense localism as a device for resolving national conundrums, then we might have to modify our assessment of Berwick.

Let us start with incomers. Berwick-upon-Tweed has a large proportion of people born and living there, and one of the lowest percentages of non-white people in England (3 per cent). According to the 2001 census, 15 per cent of the borough's population was born in Scotland (four times the proportion elsewhere in the county of Northumberland, but still less than 1 in 6). Here are some comments from those born outwith Berwick and who now live there:

We will never get to know any Berwickers because they don't want any intrusion here at all, basically, as I understand it.

They have just realised, many of them, that you don't have to brick up your windows to dodge tax!

Real Berwickers have got their own language, which we can't understand, gypsy language.

Time warp stuff. Local people, people who haven't done anything or been anywhere or whatever are just satisfied to be here.

R1 (Mrs):   It's not something that's brought out into the open. It's not something that's ever said, that you are an incomer. 'Oh, you don't come from Berwick.' But it's just the way we feel.

R2 (Mr):   It's definitely the relationships, because you go: 'Do you know so and so?', and they go: 'Oh yes. He's a cousin of my wife, and they lived at so and so, and he knows so and so'. They are very much interrelated and integrated, and as I said, to break that circle ... you can't upset anybody, because it's likely to be a relative of someone else.

Contrast those observations with these by 'locals':

R3 (Mrs):   Oh aye, we know most people, but in saying that, there's a lot of people moving into Berwick, that we don't know – interlowpers, they're ca'ed. Aren't they, Wullie? There's so many ...

R4 (Mr) (Wullie): There's proper strangers coming in ...

R3 (Mrs):   Strangers coming in every day, and we don't know them ...

Or again, slightly alarmingly:

I: How do people from Berwick get on with newcomers to the town?
R5:   OK, I would say. We've never fought wi' onybody ...

And a final example:

R6 (Mr):   They'll sort of say: 'Well, then, who's yer faither?' You'll sort of
            say: 'Jacky Hunter; he was a joiner.' 'Oh, aye, aye.' 'He worked
            wi' Bobby Higgins.' 'Aye, aye, I place you now.'
                I'm working up at the quarry, right. Now up there, there's this
            shovel driver, Colin, who lives at Burnmouth [on the Scottish
            side], and he said: 'Dougie', he says, 'I didn't realise until the other
            day that you were related to Billy Dickson, which is my mother's
            brother.' I says, 'Aye. Oh, he was merrit [married] onto my cousin.'

To be sure, it is not unusual for people living in small, self-contained
communities to constantly discuss kinship and friendship ties, and in
some places express similar sentiments, verging on distrust of incomers.[4] What is unusual is the 'national' dimension which emerges. Here
is a comment by an 'incomer' (who has lived in Berwick for more than
30 years):

They think they [Berwickers] would class themselves as more Scotch than
English, the locals, I would say so, because they talk about 'the English'.
Mrs J [neighbour] talks about 'the English people'. If someone comes from
England like I did, I think to myself – good heavens, you *are* in England
now, you know!

We have seen that many of our respondents allude to how others see
them, sometimes reacting against the stereotype of being English (or
Scots) or even Berwickers. This makes the important point that identities are often sustained or denied by significant others. What, then,
of people who do not live in Berwick but in the small villages close to
the town on both sides of the border? How far does one have to travel
from Berwick for perceptions of and claims to identity to change?

---

[4] For example, the Shetland Family History Society publishes a quarterly journal
  called *Coontin Kin* (www.shetland-fhs.org.uk/coontin-kin). The title refers to the
  almost universal Shetland custom of placing people by those to whom they are
  related; there is no suggestion in Shetland that incomers are distrusted by locals.

Let us start with the two villages on the English side of the border: Horncliffe and Norham which sit on the south bank of the Tweed and are only 3 and 4 miles respectively from Berwick. None of those we interviewed saw themselves as Berwickers, and one commented: 'I mean, you couldn't call them [people from Norham] Berwickers; they'd be totally insulted'. Another called Berwick a 'rough little town; pretty brutal', referring to it as 'English, with Scottish overtones'. And a couple had this exchange:

R1 (Mrs): Well, Berwickers are a breed of their own, aren't they really?
R2 (Mr): Aye, they're strange people, Berwickers.
R1 (Mrs): Backstabbers come to mind …
R2 (Mr): Aye, and it's no' just strangers they back-stab; they back-stab each other's families.

The consensus in the two villages when discussing Berwick was remarkably consistent:

Neither Scotland nor England; a unique place, like.

More English than Scottish, and stuck in the middle of nowhere.

Neither nowt nor something …

Just think of it as Berwick; right on the frontier.[5]

All our respondents from these two villages, without exception, used Berwick for shopping and services, almost on a daily basis, so they spoke from experience. Second, the ten people we spoke to in the local villages had no problems with claiming a national identity and considered themselves to be either English (six) or Scots (four), mainly according to place of birth. Their national identity made no difference to their perspectives on Berwick.

What of the Scottish villages? This time we spoke to ten people in Paxton and Chirnside, 4 and 8 miles from Berwick respectively, this time splitting six to four in terms of self-describing as Scottish or English. Once more, people used Berwick as the local town, so attitudes were driven by knowledge and experience. Once more, we found uniqueness and ambiguity the main descriptors:

[5] 'Frontier', as opposed to simply 'border', carries the added implication of 'beyond civilisation' (as in the American 'wild frontier').

Berwick is a special place: not England nor Scotland.

The most unique place in Britain: a sort of no-man's land.

It's more like a borders town, rather than 'English' as such.

I guess Berwick is an English town, but we never think of it like that, really.

Again we find this sense of Berwick as other, as distinct, situated for many Scots at least as 'in England' but not 'really English' (the contrast is with the small towns and villages further to the south such as Belford, Wooler and Alnwick – ineffably 'Northumbrian'). Many said they found Berwick as 'not truly English', reflected in the architecture and, above all, forms of speech and accent, somewhere between lowland Scots and northern English. One issue which exercised those in the Scottish villages was the dislocation of Berwick from its historic county, Berwickshire, reflected in what one person called 'postcode wars'. It seems that Royal Mail had decreed that Paxton was, for postal services, in 'Berwick-upon-Tweed', which meant that the correct address was 'Northumberland' – hence, England! Paxton, however, is in 'Berwickshire' – thus in Scotland, and people ran a campaign to have their mail treated accordingly as 'Scottish'. One man told us of his battle to have his roof treated as 'Scottish' so as to conform to Scottish building standards (this involves nailing slates on to what are known as sarking boards), while local builders (in Berwick) were required to conform to English standards (using timber battens). In practice, what happens, as we shall see when we look at organisational structures in Berwick, is companies and firms specialising in either English or in Scottish practices, or having an 'English' and a 'Scottish' department (such as in legal practices). Sometimes these cross-border arrangements can be Kafkaesque, as in the case of one woman in Paxton (Scotland) who wanted to start a small business. Like many others, she had no sense of 'crossing the border' when she went to and from Berwick, and most of the time it is a taken-for-granted activity. However, she said:

There are times when it's really annoying. For instance a couple of months ago we decided to start up in business, and there is a Business Initiative place in Berwick, but they said they can't help me because you live on the other side of the border. That's when you think: well, I shop here, I do everything here, my doctor, the hospital and so on – but as far as that's concerned, no. And I had to go to Eyemouth [10 miles away and of course in Scotland] to get financial help. I was going to apply for the Prince's Business Trust, the youth thing, but they have a different set of rules, dependent on which side of the border you live on. And at the time I was 28, and if I'd lived in England I could have applied for it, but the age they stop it in Scotland is 27. And I went to see them, and I said – could we not, you know – I'm only a mile or so over the border … No, no; there was no leeway. Things like that; makes you think.

Others (Scots) told us of making sure that their children were born in Edinburgh or Melrose rather than at Berwick maternity hospital at Castlehills (now closed), or further south at Ashington in Northumberland. Despite the fact that Scotland and England are part of the same (British) state, those jurisdictions relating to services, such as education, health, law and so on, make living on one side or the other of the border real in their effects, particularly since the Scottish Parliament was set up in 1999 and education and health became devolved matters firming up what was previously de facto the case.[6]

It seems, then, that people living in the villages around Berwick see it as distinct, even anomalous, a place apart. There is little sense that there is a seamless web between England and Scotland. This work on the border villages was, in many ways, undertaken as a prelude to more substantive work in the two bigger towns of Eyemouth (population 3500) which lies 8 miles north of Berwick and indubitably in cultural terms in Scotland, and Alnwick (population 8000) and 30 miles to the south. The border villages showed clearly both that the attitudes to identity tail off very sharply as one moves south or north while Berwick is perceived as distinct, anomalous and curious. What then happens as one moves still further afield to Alnwick or Eyemouth? To anticipate our argument somewhat, these are the significant other places which help to confer Berwick's identity, standing as they do as polar types of what Berwick is not, thereby highlighting its curious identity.

---

[6] Our interviews were carried out in the mid 1990s during the period of agitation for Scottish Home Rule but before the Parliament had been established.

The view from the south

Let us start with this comment from a self-confessed Berwicker about identifying with people to the south:

I: Do people in Berwick think of themselves as being Northumbrians?
R: No I don't think so. I've never classed mysel' as that. As far as I'm concerned I'm a Berwicker and that's it, like.
I: Do you know why that is? Why there isn't that sense of being Northumbrian?
R: Geographical isolation. I mean, we're 30 miles north of Alnwick and that puts us 57 miles north of Morpeth [the county town], so we are just 'Berwick'. Berwick has got its own separate identity, always has done and that is a hang-up from the past.

The surprisingly weak attachment by residents of Berwick-upon-Tweed to a Northumbrian identity is a significant finding from our research. This was particularly evident when comparing the views of people in Berwick-upon-Tweed and in Alnwick, which is also in Northumberland. We asked respondents in these towns who regarded being North-umbrian as applicable to them, how often they would use it to describe themselves. While around two-thirds of people in Berwick are pre-pared to call themselves Northumbrian, it is a somewhat 'thin' identi-fication. Thus, 38 per cent of them say they would 'hardly ever' use it, and a further 35 per cent said 'occasionally'. Only 27 per cent said they would always or usually call themselves Northumbrian. This is very different from those we interviewed in Alnwick,[7] where two-thirds said they would always or usually call themselves Northumbrian, with 19 per cent 'occasionally' and 16 per cent 'hardly ever'.

Furthermore, in terms of national identity, 'Alnwickers' were more likely to view 'Berwickers' as clearly 'Scottish' (47 per cent) than clearly 'English' (27 per cent).[8] This, as we saw previously, is in sharp contrast to Berwickers themselves, 17 per cent of whom describe themselves as 'Scottish' and 46 per cent as 'English'. In other words, people in the southern town view Berwickers as 'other', having far more in common with people north of the border. Why should that be? First of all, the

[7] All the people we interviewed in Alnwick (25) acknowledged they were 'Northumbrian'.
[8] That is, the percentage saying that 'Scottish' or 'English' was an extremely or very good description of people in Berwick.

geography matters. Commented one Alnwicker: 'I think it's very hard in Berwick, where they are, right on the border, whereas we're 30 miles down. We class ourselves as English. It's hard for them up there to know *what* they are.' And another observed:

Berwick is like no-man's land. It's always been. I mean, it's on the wrong side of the Tweed to be English anyway. We've always classed the border as the Tweed anyway, haven't you? It doesn't matter which part, when you get away up, once you cross the Tweed, you're in Scotland, aren't you really? It's a border town, Berwick, isn't it? It's a border town.

That it is a 'Scottish' town is also reflected in the accent.

I think it's a Scottish town ... just sometimes when you talk to people in Berwick, they're Scotch people. The accent, aye.

Another respondent develops the point:

Berwick has an awful lot of people talk the Scottish. I don't know if it's Scotch language or whatever, but they talk Scotch, nearly everybody there. I mean, it's a Northumbrian town, but if anybody went into it from down south ... [the accent] stops about Alnwick.

One person even claimed that they could pinpoint the accent change, placing it well north of Alnwick:

I'm still amazed that you get within 5 miles of Berwick and all the way up – Geordie – up to say 10 miles south of Berwick, you get a very gentle evolution from the harsher Geordie. But there is a dividing line 5 miles south of Berwick when it suddenly becomes Scots.

Or take this exchange between a husband and wife:

R1 (Mr): Well, I have a wee bit of difficulty with the Berwick accent. Certainly, it's more Scottish, as it were, than it is English, but it isn't what I would call a true Scottish accent.
R2 (Mrs): It is a mixed breed there, isn't it?
R1 (Mr): Mixed races ...
R2 (Mrs): Well, yes. I mean it isn't sort of all Scots accent or a Northumbrian accent; it's all jumbled up there.

This perception of accent as the key marker is commented on frequently, and there is a complex relationship between accents and identity in this northerly part of England. Respondents in Alnwick often sought to distinguish themselves from those further south, as in the observation: 'No, we're not Geordies; Geordies are Newcastle people. They live on Tyneside. We are, well, we are Northumbrians', and also in this next interesting comment relating the differences to the way of life:

I mean, any idea that Geordie – that we speak Geordie up here – is an absolute nonsense. As you notice, it's a totally different dialect ... Well, with me, the difference in dialect weighs quite heavily. It's a consciousness of a totally different way of life between industrial Tyneside and the strictly agricultural life in North Northumberland, which tends to colour one's thinking and feelings in many ways.

Because as far as most Alnwickers are concerned, Berwickers are not even English, they have little in common with them as Northumbrians still less as 'English'. It is thought to be mutual:

I: How do you think people in Berwick would view people in Alnwick? How do you think they would see you?
R1 (Mr):   Indifferently ...
R2 (Mrs):   as English ...
R1 (Mr):   I don't think they'd be particularly strong in their views about it, but we're 30 miles away, and quite honestly they couldn't ...
R2 (Mrs):   ... care less.

We will have more to say about perceived antipathies between Berwick and its adjacent towns when we turn to its Scottish neighbour, the fishing town of Eyemouth, but let us reinforce the point with this anecdote from one respondent:

R3:   I remember years ago when I was going to Seahouses [on the Northumberland coast], the 'Viking' was the dance [hall] you went to on a Saturday night, and the Berwick lot were always the rough ones; they always came down to fight with the lads from this end of the border. It was Alnwick and Berwick always, and it's still the same. You still get that now; but it's a Scottish–English thing more than an Alnwick–Berwick thing.
I: So it's not just town rivalry?
R3:   No, no, it's the accent. It's the Scottish accent, definitely.

People in Alnwick seemed less aware of Berwickers' strong sense of local identity (indeed, as we shall see shortly, unlike people from Eyemouth they claim hardly ever to have heard the term 'Berwicker'), often attributing a Scottish nationality to people from Berwick. This was in spite of the fact that all but two respondents in Alnwick knew that Berwick was in England[9] and that people from the town could therefore mobilise birth, residence, upbringing and often ancestry to support an English claim. Despite all that, Berwick is seen as 'different' by its southern neighbour, even to the point of being – counter to the facts – 'Scottish' as reflected in patterns of speech and accent.

The view from the north

If people in Alnwick think of Berwick folk as 'Scottish', is this reciprocated by those who live in Eyemouth, the fishing town just 8 miles to the north of the border somewhat closer to Berwick than Alnwick? If it is, then the claim to *Scotia Irredenta* may have something going for it after all. However, put at its simplest, people from Eyemouth, while recognising that Berwickers held a strong sense of local identity, almost universally classed people from Berwick as 'English' by reference to their place of birth, which they were in no doubt was in England. If Alnwick can be described as gently English, then Eyemouth is robustly Scottish. Here is one comment: 'Berwick is England; Eyemouth is Scotland. Yes, I think more so, me, because I was born in Berwick [at Castlehills maternity hospital]. I feel for some reason more Scottish.'

I: You would never describe yourself as English?
R1 (Mr): No! ... I never told anyone I was born in Berwick. I kept it quiet for long enough ... I'll admit it now; I never forgave my mother for that.

The issue of having children born in the nearest maternity hospital exercised quite a few Eyemouth respondents. Usually, ensuring children

[9] Two of our respondents in Alnwick did actually believe that Berwick-upon-Tweed was located north of the border in Scotland and saw Berwickers unproblematically as Scottish on the basis of this.

were born in the 'right' country, that is, in Scotland, meant considerable inconvenience in travelling time. As one couple commented:

R2 (Mrs):   You could have went to Berwick to have your children. I wouldn't have done that. We had them in Edinburgh.

I: Do you know why you did that?

R2 (Mrs):   Well, I didn't want them to be born in Berwick. I wanted them to be born in Scotland. I know that's stupid, but that's ...

R3 (Mr):   A lot of people do that when they are from here. They will not go to Berwick to have their children. They will have them anywhere in Scotland, I suppose ... Silly, but there you are.

A retired registrar confirmed this practice:

Interestingly, for a couple of years when we lived in Chirnside, I was registrar for Births, Deaths and Marriages, and I was amazed how many mothers would actually insist on having their babies in Edinburgh. Which is a long trek compared to Berwick, and, really, Berwick maternity was then a lovely place: Castlehills.

Do people in Eyemouth, then, view Berwickers as English? Here are two comments:

R4 (Mr):   I think the ones I've dealt with, they're very English; but no sort of sense of identity likes of Northumberland or anything like that. They're Berwick and England, and that's it.

I: In terms of their nationality, how do you think they would describe themselves?

R5 (Mrs):   Probably English. They have been in England for 700 years or something and it's the English school system. They have GCSEs and they have English school holidays. So I think they would have to think of themselves as English, and they are in Northumberland after all.

Having said that, among some respondents in Eyemouth there appears to be more awareness than we found in Alnwick of the ambiguous and highly local quality of Berwick as a place:

I: If I were to ask if you saw Berwick as being an English town or a Scottish town, what would you say?

R6:   Neither. It is more a border town than anything. If you ask anyone from Berwick who is a right Berwicker, they will say we are Berwickers.

I: You wouldn't regard them as Scottish?

R6: No. But I wouldn't regard them as English either. They are Berwickers to me. I still think that Berwick should belong to Scotland and that the Tweed should be the border, but that's another matter.

We also find a higher level of rivalry, even antagonism, between Eyemouthers and Berwickers. Take these comments for example:

I: Have you heard the term Berwicker?

R7 (Mrs):   Yes.

I: In what sort of context?

R7 (Mrs):   Always derogatory. 'Oh, Berwicker!' As a child if someone said you were a Berwicker, it was an insult.

I: In what context might calling someone a Berwicker come up?

R8 (Mr):   Well, if just somebody says something stupid or does something daft or makes a mistake: 'What do you expect? Bloody Berwicker!'

It seems to be reciprocated:

People from Berwick call us – what do they call us? – When you play football, you are called 'cod heids', like the fishing, you know.

And once more, we find accounts of rivalries and fights:

I: In terms of Eyemouth and Berwick, are there strong relationships between the towns?

R9: A strong hatred in some ways. One hears of traditional historic fights like the people of Berwick would come in and there would be fights in the town. You know, on a Saturday night in the clubs, and dances used to get people from Berwick come in a few years back when they used to have a dance hall in Eyemouth. It was local, not a Scottish–English thing.

That comment about this being about local rather than national rivalries is interesting, and not uncommon in small neighbouring towns.[10] What does, however, give the Eyemouth–Berwick rivalry

---

[10] Indeed, we came across people in Alnwick talking about rivalries between Alnwick and Amble, and in Eyemouth between Eyemouth and Duns, the county town of Berwickshire, with a comparable population.

added piquancy is undoubtedly the national dimension, disproportionately directed from Eyemouth to Berwick, whereas Berwickers
stress local and not national identity. If anything, being English is for
them a fact of jurisdictional life, rather than something to be proud
of, especially perhaps as their English neighbours in Alnwick do not
think they are properly English at all. Eyemouth, on the other hand,
is robustly Scottish:

I: Why do you think Eyemouth has the feel of a Scottish town?
R (Mr):   Well, we are in Scotland after all, and I mean we are born and
            bred Scots people. In fact, if we fill up a form for asking what our
            nationality is, we would put in 'Scottish'. We are very proud of the
            fact that we are Scots.

What strikes the casual visitor is how 'Scottish' and 'English' Eyemouth
and Alnwick, respectively, are. This is partly a reflection of different building styles and histories, as well as distinctive – national –
accents.[11] In Berwick, by way of contrast, anyone with an interest in
national identity walking down the main street would notice names
and references on the shops and businesses, and items for sale which
would generally be common in either Scotland or England but would
not be found in both.

    If we consider the views of our respondents from Eyemouth and
Alnwick on their own national identity, then we can understand why
such a strategy of localism as practised by Berwickers might often
be met with scepticism. People from Eyemouth, for example, claimed
unproblematically to be Scottish based on birth, residence, ancestry
and upbringing. The issue of confusing or contradictory markers did
not impact on their national identity constructions as it did in Berwick.
The same was true of Alnwick, except there, respondents claimed to be
English. It would seem to be the case, therefore, that, despite a strong
sense of localism in both communities, for people in these towns to
respond to the question, 'What is your nationality?' with answers
about one's (strong) local identity would seem bizarre. Given these
contrasting national identity attributions, classed as 'English' by those
to the north and 'Scottish' by those to the south, it is understandable

---

[11] Different speech forms across the Scottish–English border have attracted
     interest from scholars of linguistics. See, for example, Glauser (1974); and
     Watt et al. (2010).

that people from Berwick sidestep questions about national identity and seek to avoid the issue by taking refuge in claiming more unproblematically to be 'Berwickers'.

What we have seen here is a complex interplay between local and national identities, the latter being especially ambiguous, allowing a claim to 'Scottish' identity in historical and cultural terms, as well as to 'English' jurisdictionally. It is the beginning and the end of England, and in the phrases of the local enterprise board Northumberland Partnership: 'Christianity[12] began here, and England ends here.' Nevertheless, Berwick recognises itself as a periphery of a periphery of a periphery. It is in the English 'North', with all the political and cultural distance this implies (Taylor, 1993), and it is Northumberland 'north of Leeds', in the words of one respondent seeking to express Berwick's perceived peripherality from London. Finally, it is itself a remote part of Northumberland which has a Labour southern core, a Conservative area in the middle while Berwick-upon-Tweed, sitting on the fence to the last, returns a Liberal Democrat MP to Westminster.

Berwick is constructed as a place, or a set of places, in terms of its geographical, historical, political and cultural distinctiveness, in which local identity is created and maintained in counterpoint to two national identities, Scottish and English. It is a place on a real border politically and ideologically; it is marked by political and geographical marginality; and it has an activated history on which it draws to emphasise its identity. Its ambiguity is its cultural asset.

The ambiguity and distinctiveness of Berwick is in large part institutional and jurisdictional. The reader will recall the tale of the woman from Paxton who had some difficulty persuading either English or Scottish authorities to fund her business enterprise. The border is 'real' because (different) political and institutional practices are real. Here are some examples which people gave us, the first from a Berwicker:

I: Do people look south rather than north?
R1: I think it tends to go south. Because we're in Northumberland, and simple things like the Yellow Pages. You don't get the Scottish [one], and you don't get the [Scottish] Borders telephone directory; you get

---

[12] The reference is to Lindisfarne, the Holy Island some eight miles south of Berwick, settled by Celtic monks from the north and west, but treated in terms of religious heritage as the fount of (English) Christianity.

the Northumberland one ... Yea, it's the Newcastle Yellow Pages; it's the Northumberland phone book, so you don't get the Borders one unless you particularly ask for it, so those kind of things make you go south. The TV station is Newcastle. Certain parts of the estate do get Border [TV] but it depends on which way your aerial faces, but we get Tyne Tees, so we're Newcastle across to Carlisle.

And in terms of political developments relating to devolution and the Scottish Parliament:

R2:   There is a lot of Scottish people work beside us, and ye hear them talking. And was there no' an election a couple of weeks ago [the Scottish Parliament election of 1999], and I didn't really think about it until I heard people talking about it. I didn't give it a thought until I realised that they was going to try and get their own parliament.

R3:   There is a lot of people round here would prefer to be called Scotch. I think that is where there is a lot of worry is in Berwick; is what is going to happen to Berwick if Scotland does go on its own ... there is concern. There is a lot of people sort of saying: what is going to happen to us? England probably doesn't want us. And Scotland probably doesn't want us either.

Even something as seemingly mundane as the roads crossing the border matter, as one Berwick man observed:

Because I look at the road situations; the Scottish roads coming doon, and they're decent until they get to just outside the toon. They run out of money at this side coming up to Berwick; we can never get anything done on, like, the road improvements. So to me it's just a no-man's land,[13] where people dinna want you.

Roads and jurisdictions figure quite a lot in people's border talk. Here is one example from a man living in one of the (English) border villages:

---

[13] The term 'no-man's land' figures in everyday speech, and possibly derives from the long history of dispute over territory in the Scottish–English borders and their characterisation as 'debatable lands' with no settled authority. It is a popular theme in books such as George Macdonald Fraser's *The Steel Bonnets* (1971) and, more recently, Andrew Greig's novel *Fair Helen* (2013).

It's never been an issue, really, that there was a border. The only time you tend to think of the border is when you're coming back in and there is a police car behind you, and you think: well, if I can get across this border, even if he stops me for bald tyres, you know … because once you're back over and into England, he can't stop you! That just happened to me the other day, before the van went in for its MoT, and I knew one of the tyres was going to have to be changed anyway. So I saw this patrol on the Scottish side, and headed swiftly for England where they couldn't get me!

To be sure, most people we spoke to treated the border as an almost invisible fact of everyday life, crossing it to buy a pint of milk or a loaf of bread; no, they never thought about it; saw it as something for tourists; didn't usually notice the road signs; sometimes noticed that the road surfaces were different, but at the same time, spoke of little rituals they and their families had once they were back in 'God's own country' (depending on your point of view); wiping the soles of your feet; making jokes about changing air quality, and so on. Joking relationships frequently reflect meaningful social realities (Davies, 1990). Jocular references to the border are the everyday expressions of historical fact, of jurisdictional differences, of facets of political practice, particularly with Scottish and British governments operating different policies on either side of the border.

Associational life and identity

Our research on Berwick and its environs documented a number of organisations and associations, looking at how they saw themselves in identity terms, and how they managed their businesses. Some might think that the border is a cultural joke, but it matters. Take this example given by the local Church of England vicar:

It [the border] had an immediate influence when I first came here, that I actually had the registrar in Berwick ringing me on the Friday before a wedding saying: 'Oh by the way, the wedding you have on tomorrow, you can go ahead if you want but do tell them to scrub the honeymoon and come and see me on Monday, and fill in the forms because it won't be valid!' 'Oh, yes?', says I. And he replied 'Oh yes, vicar. You're doing a cross-border marriage!' And I had no idea what all this meant. And what it meant was that there is some disharmony between the registrars in Scotland and in

England so that they won't accept the calling of banns in a Scottish parish church and the calling of banns in an English parish church where one is registered, that is, one registered here and one in Scotland.

There are examples of all sorts of anomalies, some a reflection of the postcode confusions, such that TD15 is the Berwickshire postcode, but Berwick is jurisdictionally in England. Accordingly, when the local government commission surveyed 'English' opinion in the 1970s, the town was excluded because ostensibly it was 'in Scotland'.

So how do associations manage the identity question? Perhaps the most famous in sporting terms is Berwick Rangers (known as the 'wee Rangers' to distinguish them from Glasgow Rangers, whom they famously beat in the Scottish Cup in 1967) which plays in the Scottish football league, and celebrates the fact. It has to be handled carefully, as the club secretary explained:

I've said on heaps of occasions that we are the only English team in Scotland, or the only Scottish team in England, so we can swing both ways. Anyway, within the town there is a lot of hard-core supporters who fancy themselves as Englishmen in Scotland, as well as another lot that fancy themselves as Scots in England, so we have to be careful.

He spoke of the common call of opposing supporters: 'Get into those English b*******!' Mind you, he said with equanimity, 'if it wasn't anti-English, it'd be anti-something else'.

Of the 29 associations we interviewed, 9 saw themselves as English, 5 as Scottish, and the rest neither or both, which is probably a fair reflection of Berwick identity more generally among the local population. So on what basis do associations make those claims? First of all, there is the matter of which national association they choose to be affiliated to, if they have the choice. Thus, most famously, Berwick Rangers FC, but also the local bowling club, the cycling club, the Round Table (but not the Rotary – they choose to be English) are affiliated in Scotland, for the pragmatic reason that it cuts down travelling costs. 'English' associations tend to reflect local or central government funding for arts, heritage and culture. Then there are legal matters. The Berwick History Society said: 'In all practical matters, it has to be English; it's the law of the land, and the land is English.' Some are at pains to avoid the question: 'we're just a local swimming club, that's what we are';

or the ladies hockey club ('we don't like to think about it like that'); the choral society ('we don't want to upset anybody'); and the local Labour Party which proclaimed, somewhat cryptically, that 'Labour doesn't have a national identity'. In short, it is a matter of the most advantageous affiliation – usually determined by a need to maximise the chances of getting a game (such as the local cricket club – whose first team play in Scotland, and the seconds in England), of obtaining funding (especially if tied to local authorities), or simply the predilections of the office-holders ('I'm English, so that's good enough for this club' – the local harriers).

Does any of this matter? It does, if you are placed on an ambiguous border and wish to keep up a good level of civil and associational life. So too the commercial organisations such as law firms and the banks who manage to keep a foot (and staff) in both camps. 'Ah', said one, 'you want to talk to our lawyer who deals with *Scottish* matters!'

Throughout our research, we encountered the claim that Berwick was split 50/50 between England and Scotland, but that local identity was far more significant than either. In practice, we found little evidence that the citizenry of the border town was evenly split down the middle. If anything, most thought of themselves as English, but simply because that's where Berwick has ended up jurisdictionally. With hindsight, we might treat this claim not as an empirical statement relating to an individual's *own* identity so much as an identity attaching to the town itself – a collective characteristic, given its long and often conflicted history. As we mentioned earlier, the local tourist board still describes the Elizabethan walls as 'keeping out marauding Scots', as if the Scots had no historic right to be there and almost might invade again at any moment.

Let us end with a nice example of dual identities which makes the point well. The local weekly newspaper, based in Berwick and published every Tuesday, is owned by one company that issues two editions – one Scottish (*The Berwickshire News*), one English (*Berwick Advertiser*). There is very little overlap. Comparing, for example, the issues for 18 April 2014, there is one common feature: that Liberal Democrats want a 'practical approach' to cross-border health issues in the Scottish independence debate. The *Berwick Advertiser* leads on four English stories: opposition to a local wind farm, the Lindisfarne seabird centre, the development of a 'Heavy Horse' centre and Berwick's £1 million parks project. *The Berwickshire News* looks north of the border, covering a swimmer from Duns qualifying for the Commonwealth Games, a Cockburnspath school event in liaison with the (Scottish) John Muir Trust, and the Common Riding at Coldstream. Plainly, each newspaper is aimed at different, nationally focused, markets.

Does any of this matter? Surely, some might say, this is banal stuff. Indeed so, and that is important; we agree with Michael Billig that the banal, the 'taken for granted', is basic, fundamental to who we think we are and choose to be. One might take the view that all this politics of identity is 'history', but history is never quite 'dead', especially in the context of political and constitutional change. Indeed, the Scottish independence referendum of 2014 highlights these issues. For opponents of independence, it has visions of border posts, passport controls and the apparatus of crossing into 'foreign' territory. Those who cross this 'border' on a daily basis treat it as banal. On the other hand, there is a clear division between England and Scotland, and jurisdictional differences which existed long before devolution occurred in 1999, and which the Scottish Parliament has helped to amplify. It is possible that 'independence' would simply reinforce the contrary nature of this border town. Periodically Scottish nationalists (famously, Wendy Wood, but also Winnie Ewing and current MSP Christine Grahame) make a stand, literally, on the border bridge, proclaiming that old Berwick is Scottish and painting a line across the middle of it. Berwickers notice this, usually with wry amusement, and for them it probably just confirms that there is England, there is Scotland – but above all, there is Berwick-upon-Tweed.

5 | Claiming national identity

We have seen in previous chapters that national identity matters to people in England as well as Scotland. It is an important part of the way people in both societies describe themselves, although it operates somewhat differently north and south of the border. As we showed in Chapter 2, Scots, whether they live in England or Scotland, are well able to 'talk the talk', whereas, as Susan Condor and her colleagues show, identity talk in England is more muted (Condor, 2006).

Being able to show, however, that national identity matters to people in both countries is only a first step. In Chapter 2 we introduced the idea of identity markers and identity rules, that is, the characteristics which people use in claiming national identity or attributing it to others, and the rules or procedures they employ in the process. We regard these markers and rules as the conventions by which people develop and maintain a sense of their own and others' national identity. Such markers and rules are usually implicit and unproblematic, which is why we sought out contexts where national identities were problematic or ambiguous. Of course these are not 'typical', but they have heuristic value in terms of research design as they make more apparent what is normally taken for granted.

In this chapter, we will explore how people deal with claims to national identity; how they present their own claims; how they attribute national identity to others; and how they themselves receive the claims of others, or react to attributed claims by a third party. Studying these processes enables us to address a key question of societal structure: who is 'one of us'; who do we *include* as 'people like us', and who do we choose to exclude. Most of the time this matters little, but it can become a live political issue. It may influence who has access to society's resources and rights, because the less 'like us' people are judged to be, the more restricted are their social rights, especially in periods of austerity and cuts in social welfare. Indeed, the rise of the 'radical right' in many northern European countries, including Scandinavia, is a reflection of this kind of politics of identity (Meret and Siim, 2013).

We have made the point that identity rules are probabilistic rules of thumb whereby under certain conditions and in particular contexts, identity markers are interpreted, combined or given precedence over others. They are guidelines, though not definitive or unambiguous ones, to the identity markers people mobilise in their identity claims, as well as those they will look to when seeking to attribute national identity or judge the claims and attributions made by others. Such rules are generally glimpsed only in their transgression, where, for example, claims are made which are judged illegitimate. In this way, they resemble many social norms on how to behave – such as forming a queue, interacting with people in public – where only transgression makes the 'rule' explicit. Much of social behaviour is ordered in this way, which makes it difficult for the incomer to know how to behave except by watching and listening to others.

How does this relate to 'rules' of national identification? In Chapter 2 we saw that the most common marker of 'being Scottish' was birthplace, followed by ancestry and residence in Scotland. The point about such markers of national identity is that they allow us to play our hand appropriately in the contexts in which we find ourselves. As Richard Jenkins (2008: 127–8) pointed out, you cannot turn up at the Norwegian border (or any other for that matter), claiming to be 'Norwegian' if you do not have the relevant passport, or language, or ancestral or historical connection to Norway. Claims have to be based on commonly accepted, even legalistic, rules. 'I am one of you because I want to be', is rarely sufficient to let you join the national club. On the other hand, while 'blood' or ancestry is one such card to play, in and of itself it can be problematic, and the history of the twentieth century makes us leery of blood claims.

That said, having the requisite 'patriality' without actual citizenship can bestow rights and privileges. For instance, the UK Immigration Act of 1971 conferred the 'right of abode' on those with a parent or grandparent born in the UK. Place of birth is an especially strong identity marker, even though no-one has much control over this 'accident' of birth, and our parents, it is generally safe to assume, did not give it a great deal of thought. Nevertheless, as we reported in the previous chapter, there are some Berwickers who chose to have their children born in Scotland in order to make them 'Scottish'. There are also claims based on commitment and a sense of belonging, notably among those living but not born in the country. Their key markers for claiming to belong are demonstrable forms of commitment and contribution to the country. Such people usually rely upon extended residence, which enables them to acquire cultural aspects and feelings of attachment and commitment to Scotland through late rather than through early socialisation. Recognition as a Scot may result primarily from choosing to *be* one; in the words of one of our respondents, *living* the identity, even if seldom explicitly or definitively claiming it.

Talking about identity claims

In this chapter, we will see if we can clarify some of the underlying rules by examining how people go about making, and judging, claims to national identity. We begin with some instances of migrants talking about national identity. Here are some examples taken from our intensive interviews with English-born migrants to Scotland.

'I live the Scottish way'

I: Can I ask why you don't think that you can claim to be Scottish?

R: Because I'm being honest. I like being here, I like being associated with the people but I don't like to go out into the street and say 'Yeah, I'm a better Scot than you.' Because people could quite rightly turn round and say, 'You're an Englishman.'

I: So it's because of your place of birth that you can never think of yourself as being Scottish?

R: Not really, because I think it goes back to where you were born. It doesn't have to be I suppose but I think if you are here, born in Scotland and your name is Mc-something it's ten to one you'll have a tartan attached to your name and you can find lists of them. So people can go 'Oh aye, he's a McDiarmid or he's a Cameron' and you can practically then identify it with a part of Scotland also, that the name historically came from.

I: There's no sense that because you're living in Scotland, you're then Scottish?

R: Not really, I live the Scottish way, let's put it that way, but that doesn't make me a Scotsman.

I: Even if you might choose to want to make that claim?

R: (Pause) There's a possibility that could be so but to me if you were to
      cast it down in writing there's an element to it not being true. If I'm
      writing a thing down it should be factually correct.

This conversation highlights certain features of national identity: where you were born, who your ancestors were, as well as something more indefinable, but clearly important to the respondent: living
'the Scottish way', even though he judges that insufficient to make him
a Scotsman.

How do people make a 'belonging' claim through long-term commitment? Here are two examples:

'Our home is here now'

R1:   I don't think of myself as English at all. Really, I feel, as much as any-
      thing, Scottish, if you want to ... because I've been here so long now,
      thirty-five years. My wife's been here longer than she lived in England.
      Our home is here now. We're not moving from here. I'll die here and,
      as I say, three of my children are living here and I identify with it.
      The charities I contribute to, the charitable things I try to do, are for
      Scotland. Scotland's given me a lot and I owe something.

And another:

R2:   I'd probably say I was Scottish because I've lived here mostly ... I have
      a stronger sense of being Scottish but then that's a culture thing. It's
      more the things you do, the people you see, the places you're used to
      going to, to the kind of lifestyle we lead. I think the Scottish way of
      life is slightly different to the English way of life. It's much faster down
      south unless you're in a really rural area, I would say. I find we have a
      lot more personal space in Scotland.

Both respondents tread carefully in making their claims. The first uses
circumspect phrases such as '*think* of myself as', and '*feel*, as much
as anything', and while the second respondent uses slightly stronger
language ('I'd probably say I was Scottish'), they follow it up quickly
with the justification – 'because I've lived here mostly', and the second
respondent also invokes 'the Scottish way of life', an idea we have
encountered earlier in this chapter.

What then prevents people simply claiming to be Scottish? The following respondent explains to the interviewer why they could not
make the claim:

I: You wouldn't feel able to say 'I'm Scottish'?

R3:   Well, I'm not, because I was born in England. I feel Scottish but I can't
      introduce myself as Scottish because of the importance of where you're
      born ... I would feel that if I'm talking to someone that I don't know,
      they'll say 'Well you haven't got a Scottish accent so you can't be.' It's
      as if you're claiming something for yourself to make yourself socially
      acceptable. So I wouldn't tend to do that.

This person says that they 'feel Scottish', but judges that any such claim
would be contradicted by a key marker, their accent ('so you can't be'),
which in this respondent's case could not be countered by appealing to
place of birth. This person is aware that making a claim of this sort is
one thing; having it accepted is another. That raises the question: why
would Scottish nationals not accept such claims?
    Here are some examples of how Scottish 'nationals' assess the claims
of English migrants:

'... claiming to be something you're not ...'

R4:   Do you think people would want to claim to be Scottish if they
      weren't? (laughs)

I: That would be the thing, if somebody came from somewhere else,
      they'd probably wish to continue to still make the claim to wherever
      they came from originally?

R4:   Yeah, certainly if I was in that position, I would certainly claim to be
      Scottish. I don't think there's any point in claiming to be something
      that you're not but any first generation that come over here, emigrate,
      be it from wherever, I've got a feeling that as long as they try and inte-
      grate into the community, try and mix, speak our language, make an
      effort to fit in, I think that's fine. But I don't see why they would want
      to call themselves Scottish but there you go.

And another:

R5:   I think if somebody was born in England, they'd always class their-
      selves as English. Well personally, I was born in Scotland. There is no
      way I'd ever class myself as being English or British. I'm Scottish.

The first respondent (R4) treats national identity as a fact of birth,
and thus a fact of life: you either are Scottish, or you are not, based
on where you were born, and while adapting as best you can ('try

and integrate', 'try and mix', 'speak our language', 'make an effort to fit in') you can never be Scottish. And in any case, why would they want to call themselves Scottish? The second respondent (R5) assumes that, just as they would always see themselves as Scottish, someone born in England would do the same, 'they'd always class theirselves as English'.

Scottish 'nationals' tend to conceive of their own Scottish identity, and national identities more generally, as matters of *birth*, and to a much lesser extent *belonging*. Like the last respondent, they also assume that others 'do identity' in similar ways to themselves, an assumption which is in itself a vital part of imagining who is 'one of us'. As a result, it seems a priori likely that explicit claims by migrants based on belonging would be rejected by most nationals, given the emphasis which they themselves place on birth. Consider this 'national's' comment:

They can't *become* Scottish; they can be integrated into a Scottish community. I think that's fine but as far as becoming Scottish. To go back to what I said before, not that generation but then the next generation. So if their kids are born in Scotland, if they choose to become, I think that's fine.

'English John'

Nevertheless, people are aware that claims have to be judged when made in particular contexts, and hence, the rules underlying the previous examples do not always apply. Here is a good example:

I: Do you think it would be possible for someone to become Scottish?
R: Absolutely. John (pseudonym), the English guy, he's English but his father was Asian and he becomes more and more Scottish every day. He's just basically one of the boys and the only time basically it's remembered that he's English is when we play them at football and that's it, suddenly John's English again. But his accent has even changed, he's as Scottish as anyone else, because now he's lived here for this long that this is his patch as well.
I: Would he, in any way, describe himself as being Scottish?
R: He probably could do if the boys didn't remind him, at times, as I say, usually sporting times, that he was English. He probably could do. Put it this way, to his mates in Bradford, he's like Scottish John. They'll wind him up about his accent or just the way that he talks and does things because he does things Scottish, stupid things. Aye, to his

Bradford mates he's became Scottish John but to us he has became Scottish John but really he's still *English John*, I know it sounds daft but ... (pauses and laughs).

Such a clearly elaborated response conveys some of the rich texture of people's responses to issues of national identity, who is or is not considered to be 'one of us'. We wondered after we carried out these intensive interviews, whether we could devise survey questions which would allow us to identify more systematically the tipping points between rejection and acceptance of claims. Given that people thought in terms of a bundle of markers – birthplace, accent, residence, commitment and so on – which they could bring into play as circumstances required, at what point does rejecting a claim flip over into accepting it when a new marker (such as accent) is introduced? There is an obvious problem with survey data; we are seeing how people respond hypothetically in an interview, rather than observing how they actually behave in real-life situations. We accept that in real-life interactions the context will doubtless modify to some extent how people behave, and that other subtler influences are likely to come into play. Nevertheless, as we shall see, the findings are so meaningfully structured that it is highly likely that they reflect actual behaviour and experiences.

Getting at claims to national identity

To show how this was done, and what effect it has had on our understanding of national identity, let us begin with some of our early attempts to investigate claims. In Chapter 2 we outlined these questions when discussing research instruments designed to 'get at' national identity. Here, we will discuss what we found. Our qualitative work had confirmed that place of birth is the main criterion on which people's claims are judged, and that accent is important in face-to-face interaction, so we took this as our starting point. It is well known that people like to present themselves as more 'liberal' than others and their own assessments often differ from those they attribute to the population as a whole. Accordingly, we began by using a familiar technique, asking a pair of questions, one aimed at finding out what respondents thought 'most people' would say, and the other what their own personal views were. The questions were asked in England as well

as Scotland in the 2003 British and Scottish Social Attitudes Surveys.
The questions asked in England were as follows:

*I'd like you to think of someone who was born in Scotland but now lives
permanently in England and said they were English. Do you think most
people would consider them to be English? And do you think you would
consider them to be English?*

In Scotland, the question referred to a person born in England, now
living permanently in Scotland, and saying they were Scottish.

We knew already that being Scottish is more important to people
in Scotland than being English is to people in England. However, in
the event, quite contrary to what that might lead one to expect, we
discovered that there was *virtually no difference* between the Scots
and the English in terms of what they thought 'most people' would
do. In England 32 per cent think most people would accept a claim
from someone born in the 'other' country, and the equivalent figure in Scotland was 30 per cent. On the other hand, when it comes
to 'you yourself', the differences are somewhat larger and statistically significant. While 35 per cent of respondents in England would
accept such a claim, the figure in Scotland is 44 per cent. Again
contrary to what one might expect, Scots were somewhat more willing to accept the claim than the English. That said, we should not
lose sight of the fact that well under half were willing to accept
these claims; for someone born in England, just living permanently
in Scotland is by no means always sufficient to make a successful
claim to being Scottish.

Birthplace, then, is clearly an important criterion in having a claim to
national identity accepted. Accent is also an important marker of national
identity, probably because it can be read as a proxy for where people
were born and brought up. Being white or non-white is an important
and visible marker, so we also wanted to explore a different issue, that
is, what would happen if the notional person making the claim was not
white? Hence, we asked a follow-up question, first of all in England:

*And now think of a non-white person living in England who spoke with an
English accent and said they were English. Do you think most people would
consider them to be English? And do you think you would consider them
to be English?*

We then asked a Scottish variant of the question (that is, of a non-white person living in Scotland with a Scottish accent who said they were Scottish).

This time we found no significant differences at all between England and Scotland. Forty-three per cent of people in England and 42 per cent in Scotland thought 'most people' would accept such a claim; while, unsurprisingly, they thought themselves more liberal than 'most people', with 68 per cent willing to accept the claim in England and 70 per cent in Scotland. The similarities between the two countries are striking. Scotland is neither a more liberal nor a less liberal place than England on the basis of both these responses.

Let us reinforce the point with some comment from our intensive interviews, notably those in Scotland which made the point that skin colour is a marker of national identity. Consider this comment from an 'outsider', an English-born migrant living in Glasgow:

'I think it is, and it shouldn't be'

I: Do you think that being white is a very prominent marker of being Scottish?

R1: Yeah, I think it is and I think it shouldn't be. I'm not happy that it's that way. But again it's not something that you notice until you are faced with someone who isn't white. Meeting someone who is ethnically Chinese with a Scottish accent was like a revelation, because he didn't look like he sounded. And I found it quite hard, he called himself Scottish and he wore kilts and he did Scottish things and he looked Chinese.[1] And it took a bit to adjust to it. Whereas if he'd said I'm British I would have said 'yeah, fine'. But he said 'I'm Scottish' and I thought 'Oh no you're not'.

I: Almost that, I don't know, the colour of skin would be seen as contradictory to that.

R1: Yeah, but the accent was what really threw me and the kilt, that was too much … If I see a white person who says that they are Scottish I don't think about it all I just accept it. If I see someone with a different colour who says that they are Scottish I do think about it.

---

[1] An almost archetypical example, possibly the person referred to in the quote, is a man called Andy Chung who is to be found round some of the 'folk' pubs singing Scottish songs in a strong Scottish accent wearing a kilt. See www.youtube.com/watch?v=tx3CJ5fSIak (accessed 16 July 2014).

Do Scottish 'nationals', people born in Scotland, agree with this incomer?

I: Do you think that being white is still seen as a very strong indicator of being Scottish?
R2:   Well, yeah, I mean the chances are, in Glasgow, if you see a white guy, he's going to be Scottish. And the chances are, in Glasgow, if you see an Asian guy, he's going to be Scottish, probably born here, etc. ... But that doesn't mean to say that he feels Scottish.

Developing claims

These early (2003) findings on how people judge claims showed that in both England and Scotland, a permanent resident born in the 'other country', the 'wrong country' so to speak, and making the 'wrong' claim will have their claim rejected by two-thirds of respondents. So we can say that birth outweighs residence, other things being equal. Indeed, it is perhaps surprising that as many as one-third would accept the claim when the birth criterion is not met. Second, we can say that a non-white person living in England or Scotland with the requisite 'national' accent and claiming an appropriate identity will, in the absence of other markers, have the claim rejected by around one-third of our respondents and that they believe around half of people in general would do this. Despite the evidence that accent is often taken as a proxy for country of birth, it is likely that a substantial number of people, if they do not have additional firm evidence to the contrary, just *assume* that such a non-white person was not born in either England or Scotland. It is hard to think of an alternative explanation for this other than racism.

In the Scottish Social Attitudes Survey for 2005,[2] we asked a more extensive suite of questions about accepting a claim to be Scottish made by someone living in Scotland but born in England, and we varied 'race' (white or non-white) and accent (Scottish or English). Table 5.1 shows the findings.[3]

---

[2] Funding did not allow inclusion in the British Social Attitudes Survey, so we could not make comparisons between Scotland and England.
[3] Full results and detailed analysis using the four-point scale (definitely accept, probably accept, probably reject and definitely reject) are to be found in Bechhofer and McCrone (2008).

Table 5.1 *Percentages accepting and rejecting a claim to be Scottish by notional persons all living in Scotland and born in England*

	Person A: non-white with Scottish accent	Person B: white with Scottish accent	Person C: non-white with English accent	Person D: white with English accent
Accept	42	44	11	15
Reject	57	55	87	83

Note: sample size is 1549; 'Definitely would' and 'Probably would' have been combined into Accept, 'Probably would not' and 'Definitely would not' into Reject.
Source: Scottish Social Attitudes Survey, 2005.

It is clear, that 'birth' ranks above 'accent', because a minority in every column would accept this claim which is made by an English-born person irrespective of their accent or skin colour. Second, having a Scottish accent is a more powerful discriminator than 'race' because the differentials between the non-white and white persons A and B, on the one hand, and persons C and D on the other, are very small, which would not be the case if 'race' was an important factor. Similarly, by comparing person A with person C and then person B with person D, we can isolate the effect of 'accent' as a marker of Scottishness. The claim of a *non-white* person having a *Scottish* accent would be accepted by 42 per cent as opposed to 11 per cent if they had an English accent. The corresponding figures for the white person are 44 per cent and 15 per cent. Accent is a powerful marker because of its ready accessibility. Having a Scottish accent makes a significant contribution to having one's claim to be a Scot accepted, regardless of 'race'. Whether for a white or a non-white person, having a Scottish rather than English accent increases acceptance by around 30 percentage points. It is possible that this is because respondents assume that, despite being born in England, the person's Scottish accent suggests they have lived in Scotland from an early age. They might even imagine, for example, that, when they were born, the person's parents were temporarily in England and subsequently returned to Scotland.

In Scotland, it seems that birth takes precedence over accent, which in turn takes precedence over 'race'. Is England different? In 2006, we further refined and extended the suite of questions covering these markers in *both* England and Scotland. To sharpen the analysis, we asked the questions only of respondents who were 'natives', that is, people born and currently residing in the country. How do such people view the markers of national identity? They were asked:

I'd like you to think of a white person who you know was born in Scotland, but now lives permanently in England. This person says they are English. Would you consider this person to be English?

As before, they were given a card showing four possible responses plus 'Don't know'. These were: 'Definitely would'; 'Probably would'; 'Probably would not'; 'Definitely would not'. Respondents, except those who said 'Definitely would', were then asked (and offered the same choices):

What if they had an English accent? Would you consider them to be English?

Finally, excepting those who said 'Definitely would' to the previous question, they were asked:

And what if this person with an English accent also had English parents? Would you consider them to be English?

We then asked similar questions, this time asking respondents to think of a non-white person making these claims. The results were as shown in Table 5.2.[4]

The effect of successively lowering the barrier to acceptance by introducing appropriate markers is clear for whites and non-whites alike. If the only claim to English identity is permanent residence, less than half (45 per cent) would accept the claim of a white person born in Scotland. Introduce an English accent, and that rises to 60 per cent. The ability to claim English parents results in an even larger increase and we end up with four out of five people accepting the claim (81 per cent). That one in six people (17 per cent) reject even the strongest claim almost certainly reflects the importance which respondents place on birth. The figures for the hypothetical non-white person are similar, with any sizeable difference only occurring when parentage

---

[4] For full details of the analysis, see Bechhofer and McCrone (2010).

Table 5.2 *Acceptance and rejection in England of claims to be English by a person born in Scotland*

	White (%)	White with English accent (%)	White with English accent and English parents (%)	Non-white (%)	Non-white with English accent (%)	Non-white with English accent and English parents (%)
Accept	45	60	81	45	56	72
Reject	52	38	17	51	41	25

N = 2314. (Including Don't Know/Question Not Answered in each column would bring the percentages to 100%.)
Source: British Social Attitudes Survey, 2006.

Table 5.3 *Acceptance and rejection rates in Scotland*

	White (%)	White with Scottish accent (%)	White, Scottish accent and Scottish parents (%)	Non-white (%)	Non-white with Scottish accent (%)	Non-white, Scottish accent and Scottish parents (%)
Accept	44	58	81	38	50	68
Reject	55	40	18	59	48	29

N = 1302. (Including Don't Know/Question Not Answered in each column would bring the percentages to 100%.)
Source: Scottish Social Attitudes Survey, 2006.

is introduced. At that point, the claims of whites are 9 per cent more likely to be accepted than those of non-whites, all things being equal. We might conclude, then, that there appears to be some modest degree of 'racism' involved. However, these data may understate the degree of 'racism' if some people were less than truthful because they sensed the question might be tapping racism. It should also be remembered that the line between 'prejudice' and 'discrimination' is a fine one, and we cannot tell whether those who say they would not accept a non-white claim would in practice behave differently towards the person as a result. In Scotland we see the same general effect as in England (Table 5.3).

Progressively lowering the barriers again steadily increases the acceptance rate, and again the biggest leap is between the second and third columns, when parentage is introduced; a rise of 23 per cent for white claimants and rather less, 18 per cent, for non-white. There is some evidence of more prejudice against non-whites, already 8 per cent when accent is introduced and rising to 13 per cent when parentage is brought into the picture.

Despite this difference, the substantive conclusion must be that overall for both whites and non-whites the differences between the two countries in 2006 were minimal. When it comes to the claims of white persons, in no case does the rejection rate differ by more than three points, and it declines further as the barriers to acceptance are lowered. For non-white persons the differentials were larger, starting at eight points, but once more they decline steadily as indicators of national identity are added.

There is, however, a fundamental question which we addressed in subsequent surveys in 2008/2009. Are people prepared, as one would expect, to accept claims from someone who ticks just the two most basic identity boxes, birth and residence, that is to say would they accept the claim of someone born as well as living in England to be English or born and living in Scotland to be Scottish? And would they be less willing to accept the claim if the person was non-white?[5] This 'default' question (as we call it for short) led to an acceptance rate of 97 per cent in

[5] Because we also repeated the 2006 sequence in the 2008 British Social Attitudes Survey (for England) and the 2009 Scottish Social Attitudes Survey, we asked this 'default' question *before* that sequence to avoid putting any thoughts about accent or parentage into respondents' minds. The resulting analysis appears in full in Bechhofer and McCrone (2014a).

England and 99 per cent in Scotland for whites and 83 per cent and 88 per cent respectively for non-whites. We also repeated the same question sequence asked in 2006, with results which were broadly very similar.

To recap: in 2006 we had found that Scotland was marginally more 'rejectionist' than England (Tables 5.2 and 5.3), though the similarities between the two countries were vastly greater than the differences. By 2008/2009 the tendency in England to reject the *white* Scottish-born claimants with no other markers had risen by 7 percentage points to 59 per cent and the equivalent non-white figures had risen by 8 percentage points also to 59 per cent. This is compatible with a greater general awareness that 'Scots' (and remember that we generally say that birth is the crucial criterion) have a distinct identity. This interpretation is strengthened because introducing 'English accent' and then 'English parents' more or less restores parity between 2006 and 2008 for whites and non-whites alike.[6]

Looking at the *white/non-white* comparison question in England and first at the 'default', it is striking that there is a considerable level of prejudice because as many as 17 per cent reject a non-white person claiming to be English even if born and living in England. We cannot tell whether there has been a change over time because we did not ask this question in 2006. However, the other data clearly indicate that the level of prejudice does not appear to have changed. At the starting point of the full sequence in both 2006 and 2008 whites and non-whites are equally likely to be rejected in both societies. As we move across to those with an English accent and English parents, non-whites are 8 per cent more likely to be rejected in 2006 and 9 per cent more likely to be rejected in 2008.

And what of Scotland? Comparing 2006 and 2009,[7] we find only marginal differences and a very high degree of consistency as regards putative claims by *whites* and *non-whites* alike in each category,[8] so it is fair to say that there has been no change over those three years.

---

[6] For whites the rejection rate is 3 per cent higher and for non-whites 4 per cent.
[7] The relevant survey in Scotland was carried out in 2009, a year later than in England. It is unlikely to influence the results.
[8] For the first three categories – white, white with English accent, and white with English accent and English parents – the change in rates of claims rejection are, respectively, 2%, 1% and 2%. For non-whites the equivalent figures are 3%, 0% and 2%.

Looking at the *white/non-white* comparison, the 'default' question again is the striking finding. Twelve per cent reject a claim by a non-white person born and living in Scotland, albeit this is 5 per cent lower than in England. Looking at the questions asked in both years, in Scotland, unlike England, there is a 4 or 5 per cent greater tendency to reject non-whites than whites if the only information the respondent has is that the person claiming Scottish identity lives in Scotland. The gap between whites and non-whites widens in both years to 7 or 8 per cent as the marker of accent is included, rising further when parentage is introduced to 11 per cent in 2006. In 2009, however, when parentage is introduced the gap remains steady at 7 per cent. The pattern suggests no change or even a small fall in 'prejudice'.

But surely, one might say, Scots are far more 'nationalistic' than the English. Would we not therefore have expected different responses as to who is thought 'one of us'? Further, would we not also expect that people who are more 'exclusive' in national identity terms in the two countries would be more likely to reject claims? The patterns in the data are complex, and because relatively few Scots say they are 'British not Scottish' or 'British more than Scottish', direct comparisons are not possible. Broadly, in 2006, in England the more respondents said they were English,[9] the less likely they were to accept claims from white and non-white people alike. Bearing in mind the caveat about the small proportions in Scotland at the 'British' extremes of the scale, the effect of national identity is similar but rather smaller. Turning to the 2008/2009 data, the effect in England is still apparent but now rather smaller and less sharply linear while in Scotland the effect is also still apparent but smaller, and considerably smaller for whites.

That national identity does not have a stronger effect on rates of rejection of claims, especially in 2008/2009, is somewhat surprising, but we also know that other social and demographic factors play their part. Thus, statistical modelling of the 2006 data[10] shows that the admittedly modest effect of national identity persists in both England and Scotland even after controlling for variables such as social class, education, age and sex. Education, on the other hand, also had a major independent effect such that the more educated you were, especially if you had a degree, the more willing you were to accept claims.

---

[9] Measured as usual on the Moreno scale.
[10] For the full details see McCrone and Bechhofer (2010).

What conclusions can we draw about changing attitudes to national identity claims between 2006 and 2008/2009? In the first place, in 2008/2009 people in Scotland and England did not differ greatly in their willingness to accept claims made by someone resident in England or Scotland but born in the other country. This largely replicates what we found in our analysis of the 2006 data, and is the most important finding in the context of the changing political situation.

Broadly speaking, national identity, education and to a lesser extent age continue to be the main determinants as to whether people accept or reject claims, with the proviso that the interaction between age and education – older people have fewer educational qualifications simply because of their age in any case – sometimes prevents us from separating the two. It may seem counterintuitive that in Scotland, national identity is by 2009 a *less* strong determinant of accepting or rejecting claims especially from white people than in England. After all, people's personal sense of 'national' identity is much stronger in Scotland than in England. The point, of course, is that we are looking to see whether national identity discriminates in terms of *judging the claims of others*, not whether national identity *matters* to people or not.

So why has the willingness of Scots to accept white claims become apparently *less* determined by that personal sense of national identity? This is a matter for conjecture. There have been important contextual changes in England and Scotland since 2006, notably relating to political change, with a Scottish nationalist government elected in Edinburgh in 2007 (and 2011, though outwith our period of survey), and a Conservative-led government at Westminster, elected in 2010. We might speculate that, in the decade since the Scottish Parliament was established, there has been much public debate about both what it means to be Scottish, and about policies designed to foster social inclusion (Jeffery and Mitchell, 2009; Reicher et al., 2010). Perhaps such processes make people in Scotland, especially the educated, more sensitive to issues of inclusion and exclusion at least when asked for their views by an interviewer. In England, on the other hand, debates about multiculturalism, for example, have largely been concerned with *British* rather than English identity, and as we shall see in Chapter 7, who is or is not considered to be 'English' tends to be more opaque (Alibhai-Brown, 2000; Modood, 2010).

If this surmise is correct, then encouraging public debate about national identity, and what it means to be Scottish, seems to reduce the significance of national identity when it comes to accepting or rejecting claims to be considered Scottish from white persons, born in England. In England, on the other hand, national identity – considering oneself English not British – continues to be a major factor in rejecting such claims from a white or, to a lesser extent, non-white person, born in Scotland. In England, debate tends to be focused on citizenship and the importance of being British rather than on being English (Curtice and Heath, 2009). Our evidence suggests that if this leads to people feeling more British, then they are less likely to accept even very credible claims to be English (for a discussion of English identity, see Condor, 2010 and 2011).

On the other hand, Scotland and England in 2008/2009 have become much more alike on non-white claims, and this highlights the importance of education in this as in many other areas of life. In both societies, the more educated are generally less likely to reject non-white claims because 'education' has had a lot to do with reducing discrimination. This reinforces and to some extent widens beyond higher education the point that 'the experience of higher education, being exposed to a wider range of ideas and beliefs and encouraged to think critically and independently, encourages the development of liberal, tolerant views' (McCrone and Bechhofer, 2010: 937, and footnote 14).

One of us

Why is it important to set out a calculus of claims to national identity in England and Scotland, and in particular, to find the tipping points at which people are willing to switch from rejecting to accepting claims from other people as one adds markers to the equation? Our sense of who we are matters not simply to ourselves but affects the way in which we behave in the society in which we live. Defining *who* we are is a key to determining *how* we live. Let us reinforce this point with an example from research by colleagues who are social psychologists.[11] They give an account of this imaginative and revealing social experiment:

Young Scots are approached by a researcher and asked to read a passage which purportedly summarises what Scots in general take as the criteria

[11] This was reported in *Equality and Human Rights Commission Research Report 62*, '"A strong, fair and inclusive national identity": a viewpoint on the Scottish Government's Outcome 13', 2010.

for being Scottish. For some, this provides what we have called an 'ethnic definition': to be Scottish is to be born in Scotland of Scottish parents. For others, it provides a 'civic definition': to be Scottish is an act of choice and of commitment. Then, as they leave the researcher's company, they see a young woman of Chinese origin wearing a Scotland football top. She is struggling by, carrying a pile of files on top of which is perched precariously a box of pens. She stumbles, the box falls to the floor and the pens are scattered about. How do the young Scots react? In her appearance, the woman is ethnically non-Scottish, and those who have been exposed to the ethnic definition pick up relatively few pens. By her shirt, however, she is civically Scottish, and those who have been exposed to the civic definition pick up relatively more pens. (Reicher et al., 2010: 4)

From this, and other experimental work, they deduce that there are many competing ways to define identity, and that people are willing to accept different definitions; that the ways we draw the boundaries of our identity define who is or is not seen as 'one of us'; and crucially that whether we see someone as one of us, or not, does affect how we behave towards them, and whether they benefit from those common civilities which are part of everyday life.

Two examples of the 'politicisation' of cultural differences make our point that defining *who* we are is a key to determining *how* we live. The social anthropologist Thomas Eriksen has pointed out that in the Balkans war in the early 1990s, Serbs, Croats and Bosniaks were deemed to be irreconcilably and culturally incompatible even though for generations previously they had lived together amicably sharing a common identity. Eriksen observes: 'It is only when they make a difference in interaction that cultural differences are important in the creation of ethnic boundaries' (1993: 39). In other words, the sense of a 'common we' had disappeared.

Our second example is drawn from the remarkable fact that in the history of the Holocaust, not a single Jew from the lands of old Bulgaria under German control was deported to the death camps. Social psychologists Steve Reicher, Nick Hopkins and their colleagues observed that once the Jewish population was regarded as part of the national community of Bulgarians, it was self-evident that they should be protected from attack:

To the question 'should we help them?', we may or may not answer yes. However, to ask 'should we help us' seems almost absurd. Of course we should. That is what being 'us' is all about. (Reicher et al., 2010)

They explained the protection of Bulgarian Jews in terms of in-group inclusion, that there are two dimensions of social identity which impact on social solidarity: defining category boundaries in a particular way and imputing meaning to group membership. Thus 'social identities are not simply perceptions about the world as it is now but arguments intended to mobilise people to create the world as it should be in the future' (Reicher et al., 2006).

What these studies, along with our own, are pointing to is the complex grammar and syntax of national identities. It is not a matter of having fixed identities, but malleable ones which can be stretched and reinterpreted in different contexts according to diverse purposes. Consider, for example, the case of 'English John' which we discussed earlier in this chapter. For most of the time he was simply 'one of the boys', but in the context of playing football 'suddenly John's English again'. And to his Bradford mates, he becomes 'Scottish John' because of his accent, and 'because he does things Scottish'. The point is that identities are never settled, but appropriate to the context in hand. There is no 'correct answer', but multiple formulations which are in circulation at any one time. From our perspective, to point to the malleability, and subtlety, of national identity is definitely not, as some would have it, to treat it as a 'chimera' (Malesevic, 2011). Identity is neither fixed nor singular. Much depends on how the 'national we' is defined, and for what purposes, and frequently these are broadly 'political'.

When the Scottish National Party formed a minority government in the Scottish Parliament in 2007, one of its first acts was to identify fifteen 'National Outcomes' which would make Scotland 'a better place to live and a more prosperous and successful country'.[12] One of these, Outcome 13, declared: 'We take pride in a strong, fair and inclusive national identity'. When we read this, we took the view that there was something of a paradox here; our research was showing that those who expressed a strong sense of being Scottish ('Scottish not British', in the parlance) frequently adopted much tighter rules for belonging to the 'national club', so that sometimes 'strength' of national identity was at odds with 'inclusivity'. On further reflection, we can now see that the key to Outcome 13 was that it was basically

---

[12] www.scotland.gov.uk/About/Performance/scotPerforms/outcomes (accessed 9 April 2014).

a normative exercise, to instigate a debate about who is 'one of us', in other words to broaden the parameters of national inclusion. Thus, while birthplace is, as we have seen, a key taken-for-granted marker of national identity, it is by no means the only one. Birthplace may take precedence in general terms, but making a commitment to, in the words of one of our respondents, 'the Scottish way' is sometimes a valid way of making a claim. Our Leverhulme research paid particular attention to the ways the media helped to formulate and frame the 'national we' (see Kiely et al., 2006), the 'national deixis' (Petersoo, 2007). Migrants in particular, such as this English-born person, find themselves searching for media sources which 'fit' their shifting identities:

R: I'm listening more to Radio Scotland. Most of my news sources come from Radio Scotland and Guardian and Herald. Big issues, I'm conscious that I switch much more and listen to Radio Scotland a lot more and buying The Sunday Herald. Those two are two clear decisions that happened and changes. I'm listening to Radio Scotland because of the Parliament really and it's more likely to carry on.

This respondent pointed to Scottish devolution as a catalyst for this shift towards a more Scottish national frame of reference in the Scottish media. This is even more striking when he considers how independence might affect his belonging claim:

R: Ironically it [Scottish independence] would mean I would have a Scottish passport because I would be Scottish according to the last set of rules they pulled out. It's that sort of thing that I'm interested in.
I: Would that be something you'd feel quite positively about?
R: I would, yes. It would perhaps give me the legal right to say that I am Scottish, wouldn't it?

This interview took place well before the election of a Nationalist government in 2007, and a decade before the 2014 referendum debate on Scottish independence. The allusion to 'the last set of rules they pulled out' nicely captures for us the negotiated nature of national identity, that it is rule-governed. There is some irony in the possibility of 'national identity' being equated with 'citizenship' in a new political dispensation, because that raises 'civic' ways of being 'national', rather than simply 'ethnic' ones. This is how James Meek reflects vividly on the Scottish referendum of 2014, and also illustrates clearly many of the issues we have been discussing throughout this chapter:

Were Scotland to vote for independence in September, I'm certain I'd qualify for citizenship of whatever the rump Britain was called. I was born in London, I live there and my mother was born in Essex. Despite my father's Indian birth and early childhood, I'd be pushing my luck to ask for Indian citizenship. My mother's mother was Hungarian, and under that country's sweeping new citizenship law, that would be enough to get me a passport, providing I learned to speak Hungarian. As for Scotland, it, too, plans generosity for the prodigals. I'd probably qualify for Scottish citizenship on three grounds: my father's mother, Ena, was born there, of impeccable Scottish parentage; I lived there for more than ten years continuously from 1967 to 1984; I have a 'close and continuing relationship' with Scotland. My parents live there. My elder sister lives there. Some of my closest friends are there. The publisher of my novels is there. My bank account is there. I like it there. I have been fostered as a writer there in a way a Yorkshire, Cornish or Mancunian writer would not have been. Scotland has been kind to me. I'd have to apply, though. And perhaps I shall live there again, one day. (Meek, 2014: 6)

Meek goes on to point out that being a quarter Jewish, he could almost certainly obtain an Israeli passport. While this ethnicity would have earned him the label of 'crossbreed of the second degree' in the Third Reich and cost him his life if he had been deemed to have 'particularly Jewish appearance' or 'a political record that shows [he] feels and behaves like a Jew', 'Israel accordingly, under its Law of Return', offers citizenship to those with at least one Jewish grandparent. As this example graphically and chillingly shows, the markers and rules of national identity are rarely settled, and never made once and for always.

   In this chapter, we have explored how people in England and Scotland deal with claims to national identity made on the basis of a battery of markers such as birthplace, residence, accent, 'race' and parentage. Despite the consistent evidence that Scots have a much stronger sense of their national identity than the English have of theirs, we find that the similarities between the two countries clearly outweigh the differences, and that there are similar 'rules' in operation. Arguably, the most significant context in which these rules are put into practice has been political. The first decade of the twenty-first century has seen the establishment and consolidation of devolution, and notably the creation of a law-making parliament in Edinburgh, while in England there appears to be a growing debate about who is or is not 'one of us' (see Kenny, 2014). So, on the face of it, 'politics' on both sides of the border seems to frame much of the debate about national identity. This is the issue we will explore in the next chapter.

6 | The politics of national identity

> Identity only becomes an issue when it is in crisis, when something assumed to be fixed, coherent and stable is displaced by the experience of doubt and uncertainty. (Mercer, 1990: 43)

It would seem self-evident to say that national identity is 'political'. After all, in Scotland in particular the last two decades have been intensely 'political', and arguably the decade or two before that. Since the 1970s, agitation about 'home rule' has sought to mobilise national identity. The politics of national identity were brought to the fore by the rise of the Scottish National Party in the mid-1970s on the back of the discovery of North Sea oil; Labour's attempt to create a Scottish Assembly in the second half of the 1970s culminating in the first referendum in 1979; its failure in constitutional terms, and the election of the Thatcher government. The creation of, and elections to, the Scottish Parliament since 1999, especially the electoral successes of the SNP in 2007 and 2011, would seem to confirm an intimate relationship between politics and national identity. And England was not immune to the effects of these constitutional changes. Creating a Parliament in Scotland (and an Assembly in Wales) highlighted the so-called West Lothian Question (WLQ), the apparent contradiction of non-English politicians being able to 'interfere' in 'England-only' matters at Westminster while English MPs could not vote on those same issues decided at Holyrood. The topic has begun to attract the attention of political scientists such as Michael Kenny who argues that English nationhood 'remains a subject which is usually skirted rather than directly engaged, and is mainly confined to the margins of political analysis. There are still only a few academic books devoted to this question, and English national identity has been studied much less extensively than its Scottish and Welsh counterparts' (Kenny, 2014: 1).

This was deemed to have disadvantaged the English. This led, it is claimed, to a dusting off of Chesterton's famous line about 'the people of England, that has never spoken yet' and, as we said in Chapter 3, Patrick Wright's observation that this way of thinking 'remains influential to this day' (Wright, 2005). That 'the English Question' has become salient in the early twenty-first century, and that it has become more political and reactive is summed up by Robert Hazell: 'The English Question is a portmanteau heading for a whole series of questions about the government of England. These have been thrown up as a result of the Labour Government's programme of devolution to Scotland, Wales and Northern Ireland, and of regionalism in England' (Hazell, 2006: 1).

In this chapter we will explore this assertion that politics and national identity in England and Scotland are inextricably linked and reinforce each other. Patently, the claim does not imply that they are linked in similar ways. Rather, the conventional assumption is that the rise of Scottish (and Welsh) national identity has stirred the inchoate 'English Question', making the English more aware of who they are, and setting in train a developing distance between being 'English' and being 'British', to the advantage of the former. 'Speaking for England' comes more naturally to the Conservative Party than to Labour and the Liberals, and it has become a favoured trope among right-wing commentators (see, for example, Simon Heffer (1999) and Roger Scruton (1980)). Imagining England as a 'beleaguered organic community' also fits easily into the mindset of right-wing challengers to Conservative dominance such as the United Kingdom Independence Party (UKIP). Waiting for 'the English dog to bark' has become something of a journalistic pastime, one in which non-Conservative think tanks have joined. The Institute for Public Policy Research (IPPR), for example, produced a report in 2012 called *The Dog that Finally Barked: England as an Emerging Political Community*. It observed: 'It has long been predicted that devolution to Scotland, Wales and Northern Ireland would provoke an English "backlash" against the anomalies and apparent territorial inequities of a devolved UK state … there are now signs that a stirring within England is beginning to take shape' (Jones et al., 2012: 2). In particular, the report claims that being 'English' is becoming the primary source of attachment for the English (a finding ostensibly confirmed by the 2011 Census in England); that English identity has become stronger than British identity; and that there is 'strong evidence' that English identity is becoming politicised, with mainstream parties failing to take the English Question seriously. Thus, 'Despite the exhortations of successive governments that have focused exclusively on Britishness, at the popular level it is Englishness that resonates most' (Jones et al., 2012: 3).

Is this really true? And to what extent is it also the case in Scotland? Susan Condor, who is rightly critical of any simple association of constitutional questions and national identity, especially in England, has commented: 'Popular responses to the asymmetric devolution settlement appear in part to be contingent upon the tendency to frame constitutional change primarily as a matter of "national identity"' (Condor, 2012: 95). We will explore the systematic survey evidence later, but let us begin by looking at how people do national identity talk in this context.

Being Scottish

We start with the apparently more straightforward case, the Scottish one. In our previous work on people living in Scotland, both 'natives' and English-born, we have made the point that often, and notably among those who say they are 'Scottish not British', the replies are matter-of-fact, direct and non-elaborated. Thus:

Just because ...

Because I'm Scottish not British. It just makes most sense; it gives me a point of origin.

It's got to be the first one, it's got to be Scottish not British.

Just because I was born in Scotland and I'm proud of my Scottish heritage really.

In Chapter 2, we gave two examples where respondents weighed up more explicitly how to combine being Scottish and being British. They bear repeating here to make the point that while one response is almost instinctive and the other quite analytic, 'politics' does not figure in how they go about their choices:

I would say that I'm a wee bit more Scottish than British but I still feel British ... Because I am Scottish first and foremost, you know.

And more elaborately:

No. 1 (Scottish not British) is out, 'cos I do think of myself as British ... No. 3 (Equally Scottish and British) is out. I was looking at that one but I said

I was Scottish first, and I stand by that. I live here, if I'd lived in England all my life I would probably have said 'More English than British'. But if I chose 'Equally Scottish and British' it would mean I would accept a British football team, which I wouldn't, I want a Scottish one.

Contrary to the notion that national identity is profoundly 'political', none of these comments contain any reference to 'politics' or constitutional questions, and even those people who forefront 'being Scottish', whether more Scottish than British, or even Scottish not British, are not making an explicit political statement.

Not in a nationalistic way ...

Where politics *is* mentioned it is often to dismiss political and constitutional issues:

I: Do you think of yourself as being Scottish?
R1: Not in a nationalistic way, but yes, from an identity point of view, yes.
I: When you say 'from an identity point of view', I just wondered what you meant by that?
R1: I relate to Scotland; I think of myself as Scottish, but not to make myself different from anybody else. Just actually to identify who I am.

And this other respondent specifically denies that national identity is 'political':

R2: I don't think that I need that [Scottish Parliament] to make me Scottish. I feel like Scottish people are very identifiable in their own right anyway ... and it's not just an accent thing, a temperament thing. It's everything that makes them different so I don't think that you necessarily need to have political autonomy to know that. I think Scots are very confident in themselves and their sense of nationhood, I suppose.

This contrasts with the minority who put emphasis on being British. Here are two examples; in each case, the 'political' context was raised independently by the respondents themselves. Their 'Moreno' claim is in parenthesis:

As I've said, I'm quite proud of my Scottish identity, but I'm not a nationalist. (*Equally Scottish and British*)

I would tend to think of myself as British. I'm not a great believer in independence for Scotland, if that means anything ... I don't think anything in particular comes to mind of being Scottish. You see, this is why I feel that we're all really British because I don't think we're ... If you asked an Englishman what he thought about that, he probably couldn't tell you either. I feel that although I am Scottish I would rather feel more British. Why? Just because ... I'm not a separatist. (*More British than Scottish*)

There are then many ways of talking through one's national identity in Scotland, ranging from simply reasserting the category ('I'm Scottish not British') to giving some detailed account of the thought processes involved in making a decision ('I'm not that, nor that, so possibly this one etc.'). Much against what we might expect, it is those with a strong sense of being British who are more inclined to make overt political statements ('I'm not a nationalist, a separatist') rather than those who prioritise their Scottishness.

Being English

And how do the English talk about national identity? What they tell us is, in truth, not very different. Here is someone talking through at length which national identity category fits them best:

Do I *have* to tick these?

R: So which of the following statements best describes how I see myself? English not British. Well I suppose I do see myself as being British. That's actually quite a difficult question in a lot of ways because I suppose it's how I see myself so I see myself as being British but then I can't say that I'm not English. Or I can't say that I'm English and not British because by being British I'm kind of English, because there really isn't any differentiation between being British and being English outside of England.

I: Okay.

R: But I suppose I am.

I: So you can't be English not British?

R: I personally can't, I've always considered myself to be British but I do have a notion of Englishness. But my notion of Englishness is born out of my Britishness. It isn't something that is separate from my Britishness. It's kind of in there. I do speak the English language, well I try my best anyway. I've got the British passport. Can't really say English not British. I'll leave it. Oh, do I have to tick these?

I: Well, you just tick the one, or it could be that you can think of another one that might be more appropriate. Otherwise you just tick one that describes you the best.

R: So I don't tick the ones that apply.

I: No, just the one.

R: Just the one. More English than British, I don't see how I could be more English than British. That really doesn't make sense. So, equally British and equally English. To be honest I've got a similar point on this as to what I had on the other one, which was, because you're using the same notion of Englishness, if I was a black Afro-Caribbean doing this I don't know whether it would be a particularly fair question. I mean I'm having difficulty with it and I'm British, English? (laughs) I don't know what I am anymore. I think I just might emigrate and become something else. It might be a lot easier. Equally English and British, so saying I'm equally English and British would mean that I am equally both?

I: Yeah.

R: Ah, so that might be me then. So that would seem to be, because by saying I'm more British than English that's actually denying my Englishness in spite of my Britishness. Right. Oh, British not English. But I am English, but I'm only English because everybody else tells me I'm English, but I'm British.

This is a good example for our purposes because it makes the point, also supported by others we have interviewed, that by and large English people do not *confuse* 'England' and 'Britain', though this respondent also implies that *other* people, presumably non-English people, believe 'there really isn't any differentiation between being British and being English outside of England'. It also seems the case that both being 'English' and being 'British' is important to them, and that the former 'is born out of' the latter, and thus not entirely separate. All this is illuminating because the respondent spells out the decision process in some detail. Plainly, and contra Anthony Barnett (1997), it is a distinction they 'get'. Here are further examples:

Just who I am ...

R1: I suppose I associate British with, erm, with being sort of loyal and patriotic, which I'm, which I'm not particularly, (laughter) you know ... whereas English is just, is just who I am, it's just, you know, what I was born.

What does come through in the ways people talk about English national identity is how it is done in relation to 'others' in these islands – the Scots, Welsh and Irish.

It's not important to me but it is to them, so it's changing …

R2:   I suppose I'm slightly more English than British. Because the Scots go on about it, otherwise I would say I was equally, but then it's such a big deal that I think 'well if it matters so much to them maybe I am English'. It's the Scots and the Irish who push me into categorising myself as English in a sense. It's not important to me but it is to them.

Sometimes for the English, constitutional issues in Scotland and Wales do play a part:

R3:   I do write 'English'. And I think purely and simply it is in the world that we live in now, that er, you get the Welsh Assembly, er, and the Scottish Parliament er, they keep banging on, er, what they are. And er, I think there's a lot of … British people now, that you've got that, that you – you are identifying yourself er, as English rather than you used to – well, I never thought of the Welsh, the Scots, the Irish being any different. You know, we were the United Kingdom, we were all the same. But I suppose as – as time goes on that different parts of your heritage are going and you're probably desperately trying to keep an identity.

The devolution thing

I: So why do you think it's specifically now that the English are asserting themselves?

R1:   It's probably, er, because of the devolution thing, and saying … if you're really going to be Scottish, and you're really going to be Welsh, well, we might as well really, really be English. [But] I can't imagine there being an English Parliament; it already has one.

And another:

R2:   So I mean personally I wouldn't mind if we gave up the Union because I just feel English, basically.

I: You want to see an English Parliament and all that?

R2:   Hmm … well, yeah, yeah. I'd like to see an answer to the West Lothian Question because I think it's a good point that if Scotland and Wales

and Northern Ireland have their own parliaments, should their MPs
get to vote on purely English issues? And the answer really is – no. So
I would say something needs to be done about that.

And finally, a strong statement of this position:

I: Do you call yourself English rather than British?
R3:   Oh yeah. I'm not British.
I: Why aren't you British?
R3:   Well, for the simple reason there is no such thing as Britain. Scotland
      has home rule, Ireland has home rule, Wales has home rule, um,
      England has nothing, but I'm English, not British.
I: Were you British before the constitutional change?
R3:   Yeah, yeah ... and there was Great Britain. There is no longer Great
      Britain.

What such comments convey are not so much clear-cut patterns as
the nuances of national identity talk. More generally, however, do
they imply rising English identity? And how close is the connection
between Scottish national identity and the electoral successes of the
Scottish National Party? We will address these questions, first, by
exploring the extent to which people in their respective countries
think being English and Scottish is a matter of 'politics' or a matter
of 'culture'. We will then look at the reasons people give for their
identity choices. Finally, we will examine systematically the relationship between national identity, constitutional preferences and party
identification.

Is national identity 'political' or 'cultural'?

In 2006 and again in 2008/2009, we asked people in England and in
Scotland whether they thought that being English or Scottish was more
a matter of 'politics' or of 'culture'. We wanted to test out the apparently commonsensical proposition that in Scotland national identity
was likely to be more 'political'. After all, the agitation for, and the
setting up of, a parliament in 1999 seems to have been fuelled by a rise
in the proportion of people giving primacy to 'being Scottish'. Further,
conventional wisdom about England states that national identity is
less assertive, more implicit and nuanced, than in Scotland and possibly, as a result, less 'political'. In 2006 and again in 2008/2009, we

asked those who included 'national' identity (being English, Scottish) in their identity choice:[1]

Some people say that being [English/Scottish] is mainly about countryside ['landscape' in Scotland], music, [English/Scottish] sporting teams, language and literature, and so on. Others say that being [English/Scottish] is mainly about, for example, the way [England/Scotland] is governed, the parliament and how [England/Scotland] runs its affairs. Whereabouts would you put yourself on a scale between these two positions?

On the seven-point scale, '1' related to cultural matters (countryside/ landscape, music, sport, language and literature), and '7' to 'political' matters (government, parliament etc.), with five points in between. This simple distinction has the merit of forcing the issue, and it produced some counterintuitive results. We found in 2006 that both Scots and English 'natives' (i.e. people born and currently living in each country) were at the 'cultural' end of the scale, and that, if anything, Scots took a more 'cultural' and thus less 'political' view of national identity than the English.[2] Repeating the exercise in 2008/2009 produced almost identical results.[3]

Scots, of course, are far more likely than the English to give priority to 'national' over 'state' identity, and we might have expected that those describing themselves as either 'Scottish not British' or 'more Scottish than British' would take a more 'political' view than other Scots. Certainly this is not so in 2006 where the relationship between national identity and the cultural–political scale is not significant in statistical terms.[4] In 2009, the relationship is statistically significant but not straightforwardly so. Those who describe themselves as 'Scottish not British' have a mean score of 2.93, compared with the 'more Scottish than British' who score 3.31 and 'equally Scottish and British' at 3.07.[5] Looking across the four samples (2006 and 2008 in

---

[1] Hence, it was asked of those who said they were English/Scottish not British, more English/Scottish than British, equally English/Scottish and British, and more British than English/Scottish. Only those who said they were British not English/Scottish were excluded on the grounds that they did not think of themselves in 'national' terms.
[2] The mean figure in 2006 for the English sample was 3.69, and for Scots, 3.17.
[3] The mean English score in 2008 was 3.67, and for the Scots in 2009, 3.09.
[4] Analysis of variance gives an F score of 1.781, and a significance level of 0.149.
[5] The small number of people who said 'British more than Scottish' ($N = 34$) scored 2.71.

England, and 2006 and 2009 in Scotland), it would be safer to conclude that there is no systematic and clear-cut relationship between national identity and the scale in question in either country. The important, and somewhat surprising, finding on which to focus is that both the Scots and the English tend to see national identity in slightly more 'cultural' than 'political' terms, and if anything, this is more true for the Scots than it is for the English.

While respondents were willing to place themselves on the cultural–political scale, they may not regard the issue as having any great importance. Consequently, we also asked how *important* respondents felt these cultural and political matters were.[6] The English and the Scots alike saw both as important. Thus, more than two-thirds of English people put 'cultural' *and* 'political' matters in the 'very' or 'quite' important categories in both 2006 and 2008 and the figures for the Scots were almost identical (63 per cent in 2006 and 64 per cent in 2009). As before, when placing themselves on the scale from cultural to political, how people construed their national identity made little difference to these scores.

Both Scots and the English were also remarkably consistent at both time-points in rating the importance of 'culture' above 'politics'. For the Scots the score for the importance of 'culture' was 1.69 for 2006 and 1.71 for 2009, while the 'politics' score was 2.21 in both years. For the English, the comparable scores were 1.76 and 1.77 for 'culture', and 2.07 and 2.08 for 'politics'.

These findings clearly indicate that there is no support for the idea that the English take a more 'cultural', and the Scots a more 'political', view of their respective national identities; if anything, it is the other way round. Above all, it is the similarities between the two national groups which stand out.

We do not find, then, any support for the thesis that either the Scots or the English are clearly making 'political' statements about their chosen national identities. This is further confirmation of what we reported in Chapter 3, and earlier in this chapter, where we found that explicitly 'political' reasons for identity choices were the exception rather than the rule. What underlies people's choices for being 'nationals' (those saying they were only or mainly English or Scottish) is identifying with

---

[6] On a four-point scale: 1 = very important, 2 = quite important, 3 = not very important, and 4 = not important at all.

their respective national history, traditions and cultures, along with civil institutions such as education, law and community spirit (Table 3.5). It is not that 'devolution' is an unimportant factor in making people feel 'nationals', but it comes some way behind, at least in Scotland. In England, devolution (for Scotland and Wales) has to a greater extent strengthened feelings of national identity for (English) 'nationals'.

And what is the role, if any, of 'political' factors in the choice of national identity among those who say they are *equally* English/Scottish and British? We saw in Chapter 3 that it is among this group that one finds the *greatest* impact of devolution, with around half of such people ascribing their dual nationality to the fact that Scotland and Wales now have devolved institutions (see Table 3.6 for full details). Nevertheless, devolution is still the least important of the five options. There is a real sense that this group chooses dual identity because they are strongly pro-Union, presumably thinking that devolution weakens the case for independence. These 'dualists' are also more likely to agree that at some times it is more appropriate to be English/Scottish, and at other times British, and that it is not necessary to choose definitively one over the other. In both England and Scotland, they appear to be saying that they believe that national and state identities are complementary rather than competing and they are the most committed to the future of the United Kingdom as a multinational, devolved, state.

And what of the 'statists', those in England and Scotland who describe themselves as 'mainly British'? Again, as we saw in Chapter 3, there are some interesting patterns in England (there are too few – 3 per cent – Scottish Brits to allow us to make definitive statements), which again cast doubt upon easy assertions about 'being British' having a single meaning. Thus, among this group in England we come across two distinct sets of responses: one which we might characterise as 'Empire loyalist', proud of monarchy, Empire and tradition, and celebrating British achievements; and a second 'liberal' dimension, committed to including all parts of the United Kingdom and to multicultural and multinational diversity. This neat partition into 'conservative' and 'liberal' dimensions when we look for the underlying components of choice[7] does suggest that simply knowing that

[7] A short account of factor analysis which identifies underlying components is in Chapter 3. A full account of the technique as applied to national identity choices appears in Bechhofer and McCrone (2010).

someone says they are 'British' first and foremost raises more questions than it answers. Here is further evidence that simple statements by politicians about the desirability of being British, which assume that it is a straightforward matter, are naive, even misguided. (We will explore 'being British' more fully in Chapter 8.)

The English: a people who have not spoken yet?

We have now reached the point where we can examine the contention that the English have become more assertive about their 'national' as opposed to their British identity, and that this has political-constitutional consequences. First of all, let us examine the claim that there has been a significant rise in people calling themselves 'English'. We rely here once more on responses to the Moreno question. It has been asked in England in the British Social Attitudes Surveys since 1997 through to 2009. These data refer to people *resident* in England rather than only to those born there ('natives'). We have used these because it gives a longer timeline; place of birth was only asked in later British Social Attitudes Surveys.

The data shown in Table 6.1 do not support the idea of a rising trend towards 'Englishness', especially if we focus on the 'devolution years' in the decade after 1999. Thus, the proportion saying they are 'mainly English' has risen by a mere 2 per cent between 1999 and 2009, and the 'mainly British' have fallen by a similar amount over the decade.[8] Among English 'natives', that is, people born and currently living in England, essentially the same proportion, 35 percent in 2008 and 36 per cent in 2011, said they were 'mainly English'; 16 per cent and 17 per cent respectively said they were 'mainly British'. So neither English residents nor English natives show any trend towards thinking of themselves as more English.

Might it not be the case, however, that especially among the 'mainly English' there is a greater desire for 'self-government' for the English along the lines of the comments in our intensive interviews? As one of our respondents commented, 'if it's good enough for them (the Scots and the Welsh), it's good enough for us (the English)'. So is there evidence to that effect? Taking the 'devolution

---

[8] Given that 'churning' takes place, we cannot assume a straightforward shift from 'mainly British' to 'mainly English'.

Table 6.1 *'Moreno' identity choices in England, 1997–2009*
(% by column)

	1997	1999	2000	2001	2003	2007	2008	2009
English not British	7	17	18	17	17	19	16	17
More English than British	17	14	14	13	19	14	14	16
Equally English and British	45	37	34	42	31	31	41	33
More British than English	14	11	14	9	13	14	9	10
British not English	9	14	12	11	10	12	9	13
Base	3150	2718	1928	2761	1917	859	982	1940

Note: Percentages do not add up to 100 per cent because respondents who did not answer or said 'don't know' have been omitted.
Sources: British Election Study, 1997; and British Social Attitudes Surveys, 1999–2009.

decade', 1999 to 2009, we do find an increase in the proportion wanting 'England as a whole to have its own parliament with law-making powers', from 18 per cent in 1999 to 29 per cent ten years later (Ormston and Curtice, 2012). However, thinking that 'England should be governed as it is now, with laws made by the UK Parliament' remains by far the most popular option, standing at 62 per cent in 1999, and still at 49 per cent a decade later. Support for the third option, English regional assemblies, languishes at 15 per cent on both occasions.

Returning to the 'mainly English', is there any evidence that they have become stronger believers in an English Parliament?

To be sure, all those who think of themselves *in any way* as 'English' do show some movement over the decade towards preferring a specifically English Parliament, but this remains their second choice, some way behind the status quo and the shift does not vary greatly across those four categories (Table 6.2). There is also evidence from other studies (Ormston and Curtice, 2012), that devolution for Scotland has hardened opinion in England about finance for Scotland. Whereas in 2001, 20 per cent 'strongly agreed' with the view that 'now Scotland

Table 6.2 *Preferred constitutional status by national identity (England) (% by column)*

	English not British		More English than British		Equally English and British		More British than English		British not English	
	1999	2009	1999	2009	1999	2009	1999	2009	1999	2009
England governed by UK Parliament	59	41	63	44	61	53	71	52	65	65
Regional assemblies	14	15	16	13	17	17	13	18	14	13
England governed by English Parliament	23	38	19	35	18	25	11	24	17	18
Base	491	166	389	151	999	337	298	104	354	125

Notes: Percentages do not add up to 100 per cent because respondents who did not answer or said 'don't know' have been omitted.
We have used data on English 'residents' here because birthplace was not asked in 1999 and 2009, so data on 'natives' are not available.

Sources: British Social Attitudes Surveys, 1999 and 2009.

has its own parliament, it should pay for services out of taxes collected in Scotland', this had risen to 36 per cent by 2009. If we look at the 'West Lothian Question', while most people living in England do agree that 'now Scotland has its own parliament, Scottish MPs should no longer be allowed to vote in the UK House of Commons on laws that affect only England', this has risen only marginally between 2000 and 2010 (from 65 per cent to 69 per cent).

Do the English, then, think Scotland should become independent and leave the UK? In 2011, just over a quarter (26 per cent) thought this, compared with 21 per cent in 1999, but in both years, the modal preference was for a devolved Scotland within the UK (44 per cent in 1999, and 33 per cent in 2011). These figures are not significantly different from Scottish public opinion, and show neither that the English taken as a whole want the Scots to leave the UK, nor that they wish a return to the status quo ante 1999 when there was no Scottish Parliament at all. The NatCen researchers concluded: 'The modest increase in demand for an English Parliament does not appear to reflect a growing desire among those who feel English to see their national identity embodied in new national political institutions'. Perhaps, rather, as Curtice (2011) has argued, 'England is simply feeling left out of the devolution settlement' (Ormston and Curtice, 2012: 13).

This nuanced understanding of national identity in England, and its relationship with constitutional questions, is also reflected in the work of Susan Condor, who approaches the 'English Question' from a different intellectual standpoint. Condor and her fellow social psychologists, who worked with the authors of this book on the Leverhulme programme on National Identity and Constitutional Change, were responsible for the intensive interviews in the two English locations, as well as for much independent work. Condor is a critic of the view that there has been an English backlash resulting from devolution. She also argues that responses by the English to the Scottish Parliament and to devolution are more coherent and rational than survey data might suggest. Above all, she states that the English public do not base their assessment of political legitimacy (for devolution) on calculations of 'English national self-interest' defined vis-à-vis Scotland (Condor, 2010). The supposed 'failure' of an English backlash to materialise, she concludes, does not result from the failure of the English to recognise their national identity, nor does it result from the 'pathology' attributed to English character. Rather, 'in England, discourses of nationhood have historically been dissociated from values of citizenship …

Moreover, in everyday parlance, "identity" may be treated as a matter for the private sphere, and political appeals to collective identity may consequently be interpreted as a marker of self-interest rather than communitarian duty or entitlement' (Condor and Gibson, 2007: 135). With particular reference to the Scottish Parliament, she argues that the national identity frame treats political disengagement as the normatively appropriate response; in short, it is 'a matter for them [the Scots]', or, in the words of the title of one of her articles 'Everybody's entitled to their own opinion' (Condor and Gibson, 2007).

We have ourselves seen those sentiments expressed in the comments earlier in this chapter by people in England ('it's not for me to say ...'). Indeed, principled opposition to devolution and the Scottish Parliament in particular seems to emanate from what Condor calls 'liberal cosmopolitanism', which dislikes and distrusts all forms of nationalism. She concludes:

Generally, people from working-class backgrounds or with relatively low levels of formal education were liable to interpret questions about national identity as invitations to self-disclose. In contrast, people from middle-class backgrounds or with relatively high levels of education were more inclined to regard questions regarding national identity as requests for information about citizenship status. (Condor, 2011: 43)

Condor might well disagree, but it seems to us that the survey findings from the British Social Attitudes Surveys which we have reviewed above and the analysis and comments by Condor are not out of line with each other, even though they are derived from different disciplines and from different methodologies. The survey data are in our judgement, *pace* Condor, both coherent and rational. Both point to the complex and nuanced understanding of the relationship between national identity and constitutional change among English people, especially where this relates to Scotland in particular. It is not a case of the English 'wanting to have what the Scots are having',[9] but of

---

[9] The phrase 'I'll have what she's having' is attributed to the Hollywood romantic comedy *When Harry Met Sally* (1989) in Katz's deli. It has been picked up as *mots justes* as in the title for this RSA event (www.thersa. org/events/video/archive/mark-earls-and-alex-bentley), and the title of a book of the same name and by the same authors (*I'll Have What She's Having: Mapping Social Behaviour*, A. Bentley et al., MIT, 2011).

making a normatively appropriate response which in certain circumstances requires political disengagement.

Is Scotland different?

This takes us to the case of Scotland, where, given the
political-constitutional changes in the past decade or so, we might
expect to find a quite different relationship between expressions of
national identity and 'politics'.

Let us remind ourselves of trends in national identity in Scotland,
this time based on residents, so that we can compare Scotland and
England directly (see previous Table 6.2).

On the basis of the data in Table 6.3, one could not claim that
Scottish national identity had strengthened significantly since the parliament was established in 1999. Arguably this is because of the 'ceiling
effect',[10] that around two-thirds of people resident in Scotland already
thought of themselves as 'mainly Scottish' at the start of the process.
Indeed, the shift to 'being Scottish' seems to have occurred *prior* to
devolution, suggesting that it had helped to stimulate the demand for
the parliament in the first place. Looking first, as we did in England, at
those *resident* in Scotland, nearly six in every ten say they are 'mainly
Scottish' (58 per cent in 2009) and only one in fourteen that they are
mainly British (only 7 per cent in 2009). The equivalent figures for
'natives' are 67 per cent and 3 per cent. People in Scotland are clearly
more Scottish than people in England are English.

If, as we did in the case of the English, we look directly at the question
of the relationship between national identity and constitutional preferences, what evidence is there that those who say they are 'Scottish not
British', for example, have hardened their support for independence?
Could this explain the electoral success of the Scottish National Party
in 2007, and, above all, in 2011, when they won an overall majority?
While the aggregate figures for the Moreno national identity measure have remained much the same, have they over time come more
into alignment with constitutional preferences? And are Scots more
likely to say they are 'Scottish not British' at times of Scottish rather

---

[10] The idea of a ceiling effect in quantitative social science is that once the
proportion of people taking a particular view reaches a certain level, it is
unlikely to rise much further. Clearly this is a matter of judgement and varies
from example to example.

Table 6.3 *'Moreno' identity choices in Scotland, 1997–2011*
(% by column)

	1997	1999	2000	2001	2003	2007	2009	2011
Scottish not British	23	32	37	36	31	27	27	31
More Scottish than British	38	35	31	30	34	30	31	34
Equally Scottish and British	27	22	21	24	22	28	26	24
More British than Scottish	4	3	3	3	4	5	4	4
British not Scottish	4	4	4	3	4	6	4	3
Base	882	1482	1663	1605	1508	1508	1482	1197

Note: Percentages do not add up to 100 per cent because respondents who did not answer or said 'don't know' have been omitted.
Sources: Scottish Social Attitudes Surveys, 1997–2011.

than British elections, something we might expect to happen given the strongly 'Scottish' nature of elections for the Scottish Parliament? In Table 6.4 we focus on elections for Westminster and for Holyrood between 1997 and 2011. The first column gives the percentage of Scottish natives (people born and currently living in Scotland) who say they are Scottish not British. The second column shows the percentage of the entire sample who are in favour of independence; and the third column, the percentage of the 'only Scots' (column 1) who are in favour of independence.

These data indicate 'trendless fluctuation', namely that there is neither a systematic trend over time for Scots to say they are 'Scottish not British', nor a trend towards supporting independence, either by Scots as a whole or by 'exclusive Scots'. Nor do 'Scottish' elections encourage people to say they are exclusive Scots. At Scottish elections over the period, 32 per cent do this, compared with 30.5 per cent at British elections. Furthermore, the mean percentage of exclusive Scots supporting independence at Scottish and British elections is identical – 48 per cent. Even the most successful election in SNP history – 2011 – does

Table 6.4 *Scottish identity and constitutional preferences*

	% 'Scottish not British'	% pro-independence	% of 'Scottish not British' who are pro-independence
1997 (UK election)	24	28	48
1999 (Scottish election)	34	28	46
2001 (UK election)	36	27	41
2003 (Scottish election)	35	27	46
2005 (UK election)	33	38	57
2007 (Scottish election)	28	25	48
2010 (UK election)	29	25	46
2011 (Scottish election)	32	32	53

Note: based on respondents born and living in Scotland, i.e. 'natives'.

not seem to have been based on a surge in support for independence, nor on a rise in those claiming to be Scots not British.

This becomes even more clear if we look at the far from perfect associations between national identity, constitutional preferences and support for the Nationalist political party. Using data from the 2011 Scottish Social Attitudes Survey, which followed upon the SNP's victory, we are able to say that:

- Less than half (46 per cent) of those saying they were 'Scottish not British' ('exclusive Scots') identified with the SNP and only 40 per cent of people who identified with the SNP were 'exclusive Scots'.
- Only 55 per cent of SNP identifiers supported independence which is, of course, party policy, while 45 per cent preferred devolution.
- Only 62 per cent of people who supported independence identified with the SNP, while 21 per cent identified with Labour; and 15 per cent with no political party.
- Of those who supported independence, 55 per cent were 'Scottish not British' (and 37 per cent 'more Scottish than British'); just over half (54 per cent) of these 'exclusive Scots' preferred independence.

Table 6.5 *Scottish identity and SNP voting*

	% within each identity category voting SNP	
	2007	2011
Scottish not British	60	64
More Scottish than British	41	66
Equally Scottish and British	22	29
British more than Scottish and British not Scottish	9	21

The very loose relationships, which these data reveal, between national identity, constitutional preferences and party identification are in line with previous surveys and the material we presented earlier. We might possibly say that national identity is a necessary but by no means sufficient explanation for the political success of the SNP in 2011. The SNP took virtually the same share of the vote among those who said they were 'more Scottish than British' (66 per cent) as they did among the 'Scottish not British' (64 per cent). Only among those who say they are 'equally Scottish and British' does support for the Unionist parties increase, notably for Labour (47 per cent, compared with only 29 per cent for the SNP), and the trend continues among the small proportion of 'Brits' with support for the SNP declining further (to 21 per cent, compared with 32 per cent each for Labour and the Conservatives).

There is further confirmation that the SNP did especially well in 2011, not just among 'exclusive Scots' but disproportionately among 'mainly Scots' when we compare their vote share in 2007 and 2011 in national identity terms (Table 6.5).

The success of the SNP in increasing its share of the vote over the two elections by around 13 percentage points involved capturing a much higher share than previously of the 'More Scottish than British' group together with a modest gain in the 'Equally Scottish and British' group.

Resisting the 'obvious': national identity and politics

In this chapter, we have examined the relationship between national identity and 'politics', and in particular the extent to which the Scots and the English have become more 'national' as a result of the creation of a parliament in Scotland and a national assembly in Wales. Given political-constitutional developments over the last decade or so, and especially in the context of the 2014 referendum on Scottish independence, it might initially have seemed 'obvious' that the process has been fuelled by changing configurations of national identity. Nevertheless, the seemingly 'obvious' may in fact be false, as Lazarsfeld argued more than half a century ago (Lazarsfeld, 1949).

Whether we look at party affiliation or at constitutional preferences, we do not find a strong, straightforward relationship with national identity, still less one associated with feeling strongly English or Scottish. And contrary to conventional wisdom, the connection between devolution and national identity seems stronger among those who think of themselves as British, than it does among those who forefront their 'national', English or Scottish, identities. For such 'nationalists', it is pride in historical and institutional matters which is significant, and, for them, identity is not another way of expressing a desire for greater autonomy. Further, for those who think of themselves first and foremost as British, there is no simple meaning attached to that label, especially in England. In terms of constitutional preferences, we again find no clear relationship between national identity and how people wish to be governed. In England, 'nationalists' have no great desire to have an exclusively 'English' Parliament, while in Scotland, the political successes of the Nationalist party, the SNP, are not to be explained by an upsurge in Scottish national identity, nor by support for independence.

Contrary to the 'obvious' expectation that national identity and 'politics' are intimately connected, we have shown in this chapter that it is the lack of such a connection which is striking. We will return to the question of 'the British' in the penultimate chapter of the book, and explore this attenuated relationship further. First we wish to explore a key question which is central to the issue of national identity and the way we have presented it in this book: the question of the notional 'other', and we do so in the next chapter.

All forms of social identity involve 'othering'; the positing of a notional other against whom one compares the nature and strength of one's own identity. Often this involves ruling out who you are not. Of course people do not go about their daily interactions continually doing this explicitly but, as we shall see, where national identity is concerned they are aware of the process. As Reicher and Hopkins observe: 'For Scottish identity, as indeed for any social identity, who we are depends upon who we compare ourselves to' (2001: 30). It is a question of the vis-à-vis, a term which has no simple equivalent in the English language – 'against' is both far too strong and implies opposition or hostility. Most of the time, the 'other' is taken for granted and the comparisons are implicit rather than explicit, especially where there are imbalances of power or size. Thus, men are less likely than women to define themselves vis-à-vis the other sex; middle-class people vis-à-vis the working class, and so on. In terms of national identity, it would seem likely that the Scots, and the Welsh, and possibly the Irish, are intuitively presumed to define themselves vis-à-vis the English, and it similarly seems reasonable to imagine that the latter might define themselves vis-à-vis the French or Germans on the grounds that the 'Celtic' minnows are too small to be the salient and prime 'other'. Only in the last few years have researchers rigorously investigated the processes of national identification in these islands rather than simply relying on such assumptions. One of the later tasks in this chapter will be to discuss the questions in the 2008/2009 British and Scottish Social Attitudes Surveys which we asked in order to address these issues.

First, we will focus on ethnicity and national identity, because these processes are among the most significant forms of othering, something which stood out in our interview material and discussions about national identity. The term 'ethnicity' needs clarification. In everyday speech it is treated as a synonym for 'race' or skin colour. The term only made an appearance in the English language in the mid twentieth century, and entered the Oxford English Dictionary as late as the 1970s. This may seem hard to believe, given that ethnic divisions plainly existed long before. In terms of social science (see Banton, 1983), the topic seems first to have emerged – as ethnic group – in American studies of cities rather than, as we might have expected, in the 1920s and 1930s when the Chicago School was at its height. Robert Park's classic essay *The City* (Park and Burgess, 1984 [1925]), for example, has no indexed reference to 'ethnicity'. The etymology of 'ethnic' is ancient Greek: *ethnikos*, meaning pagan or heathen. *Ta ethne* meant 'foreigners', while the Latin term *natio* referred to a breed, a stock or race, usually foreigners rather than Roman citizens who were deemed to be the 'civilised' people.

The point of this brief discussion is to emphasise that our contemporary distinction between 'ethnic' and 'national' groups is a reflection of how we *choose* to see our social world, rather than a hard and fast distinction. This is how the social anthropologist Michael Banton makes the point: 'the English, Welsh and Scots are accounted nations comprising the British state, not as ethnic divisions of a British nation or race' (1983: 64). The key term is the verb 'accounted', for Banton is observing that the distinction is one of custom and usage which also reflects power relations, rather than being inherent in the terms 'ethnic' and 'national' themselves. Custom and usage are the key here, because it is rare to find the Scots described as a 'national minority' in the United Kingdom. They can be considered a 'nation', and in a UK context, a numerical 'minority', but the term 'national minority' is conspicuous by its absence in political and cultural debates. This is because, as Will Kymlicka points out:

the term 'minority' has different connotations across the West. In any event, in none of these countries was there widespread reference to general principles about 'the rights of national minorities'. Consider debates about Scots in the UK, or about Catalans in Spain, or about Slovenes in Austria. (Kymlicka, 2006: 39)

Crucially, cultural differences such as language, religion and skin colour are not primary and definitional characteristics, but are social signifiers which are the outcome, the product, of social struggles (Banton, 1983: 28). They are cultural markers which are imbued with social significance and cultural power, in the same way as the 'national' signifiers of being Scottish or English draw upon a range of such markers, for example, birthplace, accent and ancestry. There is nothing inherent in the markers as such; they operate as 'habitus' (Bourdieu, 1984), a set of everyday social practices and understandings which come to be internalised and hence 'objectified'. Thus, 'objectively regulated and regular without being in any way the product of obedience to rules, they can be collectively orchestrated without being the product of the organizing action of the conductor' (Bourdieu, 1984: 72).

As Banton points out, physical differences do not in and of themselves give rise to cultural differences. What he calls 'phenotypes', markers and signs 'read off' the physical characteristic, are the key. This is inherently a social process. In the words of Fredrik Barth:

To the extent that actors use ethnic identities to categorise themselves and others for purposes of interaction, they form ethnic groups in this organizational sense ... and some cultural factors are used by the actors as signals and emblems of differences, others are ignored, and in some relationships radical differences are played down and denied. (1981: 202–3)

The point made by Barth, Eriksen and others is that it is not necessarily 'objective' and long-standing differences, but those which seem salient to actors at a particular time and in a particular context that are the key to the way markers are mobilised. In analysing conflict in the Balkans in the early 1990s, Eriksen observes that:

Ethnic boundaries, dormant for decades, were activated; presumed cultural differences which had been irrelevant for two generations were suddenly 'remembered' and invoked as proof that it was impossible for the two groups [Bosnians and Serbs] to live side by side. (1993: 39)

He goes on to say that only when they have an impact on social behaviour do cultural differences create ethnic boundaries. Constructing 'difference' is but the first step towards social action. We take for granted, for example, that 'Scots' or 'the English' (or the 'British' for that matter) are meaningful social categories, and 'actors' in their own right, even though what we may mean by such terms is not predetermined. They only obtain cultural meaning in the processes of interaction or description, which state or imply that 'we' are different from 'them', and in a specified way. Stereotypes help to short-circuit the process: 'Scots are mean', 'The English are snobbish', 'Americans are loud' and so on, even though such assertions are not arrived at by carefully interrogating the evidence and indeed none may exist. They are shorthands, 'collectively orchestrated' in Bourdieu's words, routinised shortcuts to focus perceptions and, ultimately, action.

In these islands, they are useful, if sometimes misleading, short-cuts because they lay out some of the parameters and contexts for political action. Ethnic and national identities in the UK are infused with particular – and peculiar – meanings. 'Citizenship' came late to the UK. Prior to 1948, inhabitants of the British Isles and the British Empire were formally 'subjects' of the Crown, and UK passports until recently described the passport holder as 'British subject: Citizen of the United Kingdom and Colonies'. This practice was called into question by independent (Commonwealth) states such as Canada and India who sought to redefine 'citizenship' for immigration purposes (Goulbourne, 1991). At the time, both the main British political parties, Labour and Conservatives, for different reasons held on to multinational, non-ethnic definitions of being 'British'. This led the British to have what, in Robin Cohen's nice phrase, were 'fuzzy frontiers'.

> British identity shows a general pattern of fragmentation. Multiple axes of identification have meant that Irish, Scots, Welsh and English people, those from the white, black and brown Commonwealth, Americans, English-speakers, Europeans and even 'aliens' have had their lives intersect one with another in overlapping and complex circles of identity construction and rejection. The shape and edges of British identity are thus historically changing, often vague and, to a degree, malleable – an aspect of the British identity I have called 'a fuzzy frontier'. (Cohen, 1994: 35)

These political and cultural meanings have become an accretion on social practice without the people who utilise them being explicitly aware of their variability and diversity. There are, however, particular contexts where they become problematic and contestable. We have already discussed two such in this book – 'debatable lands' and 'debatable people'. Being accepted as 'one of us' is a matter for the judgement of others, as we saw in Chapter 5 where we analysed 'identity claims'. Definitions are not absolute but serve a particular purpose in specific situations. Banton gives the curious and amusing example of residents in north-west Ontario in Canada whom the locals called 'Ukrainians', or 'Ukes' for short, a category lumping together with the most numerous group, Ukrainians, Poles, Russians, Yugoslavs and Romanians. One man answering to 'Uke', when asked which part of the Ukraine his family came from, replied: 'They didn't. They came from Poland. I'm a Polack.' Banton observed: 'The man was willing to be taken for a Ukrainian because that was the local convention. The local people were not interested in what they saw as the finer details of differentiation in a faraway land. It is in this way that ethnic identities are redefined in new situations' (Banton, 1999: 9). He comments: 'People give to themselves names which show who they claim to be rather than who they actually are' (Banton, 1999: 15).

The vis-à-vis relationship

If social identities are constructed around being and not being then we would expect to find such talk around national identities. If 'context' is the key, then the 'devolution decade' beginning in 1999 provides an obvious example within which such national identity talk is likely to take place and have increased. Here are some examples from the interviews in England:

Because I'm not ...

I: Would you call yourself British? English?
R1: I'm English.
I: Um hum. Why, just out of interest?
R1: Because I'm not Irish, Welsh or Scottish.

And another:

I: Yeah. What would you say your nationality was?
R2: Well, I'd have to say I'm English.
I: Mm.
R2: Not British, 'cos I'm, I live in Britain, but Britain's Scotland, Wales, Northern Ireland, England, so I'd have to say I'm English when it come down to nationality.
I: Why, that's interesting. Why not British?

146 The notional other: ethnicity and national identity

R2:   I don't know, it's just I consider England as separate from Wales, and Scotland. And Ireland especially 'cos you have to get a boat to Ireland, so (laughter) ...

I: Do you feel as though you've got anything in common with, erm, people from Scotland or Wales or ...

R2:   No. They've got their own language, so, how can you have, er, well, the Scottish got their own language, I'm sure they used to, used to speak Gaelic, don't they?

In the case of the first respondent (R1), self-definition is done entirely in terms of not being 'the other' ('because I'm not etc.'). The second respondent (R2) alludes to the 'nesting' of identities – 'Britain's Scotland, Wales, Northern Ireland, England' – and then defines 'the other' by attributing language differences ('they've got their own language') to the Scots in particular. That this language difference does not exist illustrates how difference can be wrongly attributed yet have real consequences.

A third respondent also uses the 'I'm English because I'm not ...' distinction, and reinforces it by claiming that the non-English 'others' don't think of themselves as British, so why should he as an English person:

I: What is your nationality?

R3:   I'd say English.

I: English? Why English?

R3:   Well, saying that, British is still fine but there's two things. If you're Scottish, you're Scottish; if you're Welsh, you're Welsh; Irish, Irish. But if you ask most Scotsmen they're not British, they're Scottish.

I: Tell me about it.

R3:   Yeah. Definitely. If you ask a Welshman, mostly, well not so much with the Welsh but definitely the Irish, as well, they're definitely Irish, not British. And that, in a way, annoys me a bit because therefore I'm English. I mean like, for instance, football, if it's an English team, I support England. I support other teams as well, but as long as they're not playing England.

He reinforces the point subsequently by saying that, while he, as an English person, would support the other British teams ('as long as they're not playing England'), *they* would not reciprocate.

They're different from us ...

Finally, here is another nice example of imputed 'ethnic' difference which arises in a discussion between two respondents (R4 and R5):

R4: They, I mean they – they're not kind of, I mean they're broad-minded, intelligent Scottish people but there is a distinction because they are Scottish, and we are – we are English.

I: Yeah.

R4: You know, although we speak the same language. It's like the Irish isn't it? Well exactly, you know, they –

R5: They're more Celts aren't they?

R4: I think that is fair.

R5: I know we're all muddled up and everything but I mean they were Celts, we were Saxons. And they were a very brave race and they were different from us.

The point about these claims is that they are used to justify a position, regardless of the accuracy of the imputed linguistic and racial differences. These respondents have no difficulty distinguishing the English from the non-English, even to the point of inventing the differences. Neither do they appear contestatory (Scots are described as 'broad-minded', 'intelligent'); these too are taken simply as matters of fact.

What about the Scots? How do they talk about the notional others? Here are some examples, not dissimilar to those above, but with added tension over 'being taken for' English:

We are different ...

I: Do you know why? (pause) What gives you that sense of being Scottish?

R1: I don't know. I think we are ... I think we're different. I think the Irish, the Scots, the Welsh and the English, we are different. I think the Scottish people are quite down-to-earth. I think ... then again, I don't know whether we're that much different ... I don't know, really.

I: But you feel there is some sense of difference between – you mentioned, Welsh, Scots, English and Irish?

R1: I can't really put it into words, I think, I'll end up being quite rude, I think the Scottish people are a bit different, or they're different from us, we're different from them. I think there is quite a north–south divide. I think the division of the people is probably ... I think northern English people and Scottish people are quite similar.

There is considerable caution about the claim to be different, a desire not to give offence ('I'll end up being quite rude'), and reference to some imputed similarities ('I think northern English people and Scottish people are quite similar').

Justification for a strong claim to being Scottish sometimes comes in being 'taken for' English:

R2:   There's certain things that are Scottish. I don't know if it's something to do with the blood system, I don't know, but it niggles a bit when you go abroad and they'll say 'Oh you're English' and you automatically say 'I'm not English, I'm Scottish.' You can just feel your hackles getting up, 'How dare you call me English, I've got a right broad Scots accent.' Then you'd get into, not an argument, it's like a wee heated discussion. The last time we were abroad, they said 'You're British then?' I'd go 'No, we're Scottish.' 'What does your passport say?' I got so fed up I said 'It says Scotland.'

It's a Celtic thing ...

In this example, we again get difference reinforced by 'inherent' characteristics of an ethnic sort:

I: You mentioned Scottish culture, do you feel a sense of there being quite a distinct Scottish culture?
R3:   I think there is. There's a distinction in the United Kingdom between the Celtic cultures and the Anglo-Saxon cultures. They do have different heritages. They have, even now, I think, we have different outlooks on life, largely.

The point of these Scottish and English accounts is that they are being used by respondents to make the point that the Scots and the English are 'different', even to the extent of couching this difference in 'ethnic' terms, bordering on the 'racial' in some cases. This use of racialised explanation is used to underpin or underscore 'real' difference, as the following respondent (who describes himself as 'more British than Scottish', and spoke elsewhere of his dislike of Scottish nationalism) makes his point:

I: It might seem an unusual question to ask but, for yourself, what is it that gives you that sense of being Scottish?
R4:   Yeah, I can't really explain that. It's a Celtic thing, isn't it? It comes back to what I said earlier about the fact that this, being a Celt, seems

to be something that's just in your soul, if you like, if there is such a thing. You just feel … if you like, we're more excitable about … it's a pity, it's a bit nationalistic really (laughs) but it's something that's very difficult to put a finger on it. I suppose we're quite proud, very proud to be Scottish as well because we've achieved a helluva lot really.

All a bit squished nowadays …

The notion that national differences have racial undertones is elaborated in the following interchange between the interviewer and two respondents (R5 and R6). R5 was the respondent who said in a previous extract above: 'they were Celts, we were Saxons' but now qualifies this:

R5: Well, no-one's really purely English are they? I doubt anyone could say they were really English. Um, I think I say British because none of us are true-blue English. I always thought of myself, 'I'm *English*' but then when I looked into my background, I'm Irish, Spanish, Welsh, and 'I'm English, I'm English', I'm *not*; if they were coloured, I'd be multicoloured.

I: Hmm.

R5: So I can't say I'm English. No.

I: So you see English as pure, do you?

R5: Yeah, but no, I don't think there's any of them.

I: Do you know anyone who you would say is definitely English?

R5: No, because the Queen and that lot are bloody German, aren't they (laughs). No. Do we know any real English people? No. Probably not.

R6: I'd have to say I'm British because I'm obviously not pure-blooded and then after I found out Roger was half-Irish, and was bloody with him for years and he said 'I'm – there's no Irish in me!' And he was.

I: Do you have to be pure-blooded to be English do you think? Is that a …

R5: No, no.

R6: I don't, no, not really.

R5: No, because most people are – you know.

R6: I think we're all a bit – all a bit squished about that nowadays.

How can *you* be English?

Despite this 'squishing' effect, we gave, in Chapter 3, the example of a non-white person (of Pakistani origin) being unwilling to say they are English. To abbreviate, she observed that: 'I wouldn't describe myself as being English because, to me, English means being white …

to me, English means being a native of England, having generations of families previous, who have lived in England, rather than myself, like I'm just a second generation, actually I'm probably first-generation British.'

We find this trope more widely used. For example, the novelist Martin Amis caused a degree of controversy by claiming in a BBC interview[1] that:

The great thing about America is that it's an immigrant society and a Pakistani in Boston can say 'I'm an American' and all he is doing is stating the obvious ... But a Pakistani in Preston who says 'I'm an Englishman' – that statement would raise eyebrows, for the reason that there's meant to be another layer of being English. There are other qualifications, other than being a citizen of the country, and it has to do with white skin and the habits of what is regarded to be civilised society, and recognisable bourgeois society.

To his critics, Amis was being racist, while he would doubtless argue in his defence that he was simply recognising the distinction between nation and state (all Americans and America) and alluding to the association between being English and being white though the further comment about civilised society is harder to justify. As Fenton and Mann conclude from their research in south-west England, 'with respect to the distinction between Englishness and Britishness, whiteness appears important' (Fenton and Mann, 2013: 226–7). The obverse is a familiar theme encountered among non-white people in England, that they are 'British', rather than 'English'. For example, Curtice and Heath argue that as a result of British national identity having a less exclusive character than English identity, 'those who belong to an ethnic minority may be reluctant to claim an English identity, while those who do adhere to that identity may have a more restrictive view about who has the right to claim a particular national identity' (2009: 57). The issue of whether non-white people think of themselves as 'English' or 'British' has been given political salience ever since 1990 when the Conservative politician Norman Tebbit invoked the 'cricket test' whereby, he asserted, people of

---

[1] www.theguardian.com/books/2014/mar/17/martin-amis-white-skin-english-attribute-multiculturalism (accessed 20 March 2014).

Asian or West Indian extraction were less likely to support England than their countries of origin. Subsequently, the sociologist Stuart Hall observed: 'I was puzzled when Norman Tebbit asked which cricket team you would support, in order to discover whether you were "one of us", "one of them" or maybe neither. My own response to that was, if you can tell me how many of the four hundred members of the British athletics team are properly British, I'd be ready to answer the question about the cricket team; otherwise not' (Hall, 1995: 4).[2]

Taking our respondents' observations together, what do they tell us about the ways people speak about national identity? First of all, the comments carry a strong sense of the vis-à-vis; national identity cannot be taken in isolation. English people, for example, point to the fact that the non-English assert being Scottish/Welsh/Irish over or against British, so why, as English people, shouldn't they? Scots allude to being 'taken for English' when abroad, and react against it. The frequently used 'melting pot' metaphor sits alongside the view that 'political' differences between the English and non-English must derive from deep-seated differences. People search for the underlying reasons for presumed 'difference' – hence, the willingness to 'racialise' identities in some cases, to claim that there are deep, ancestral roots to all of this. This was perfectly, if perhaps not entirely consciously expressed by the respondent we quoted earlier in Chapter 3, who distinguished English from British by saying: 'English is perhaps more genetic, isn't it?'

Islamophobia and Anglophobia: the cases of Scotland and England

We can see, then, that for some people, when considering perceived national differences – English, Scottish, vis-à-vis being British – there is an issue of 'race' and ethnicity in the narrow sense of the term ('genetic', as the last respondent commented). We saw in Chapter 5 that non-white persons stand a lower chance of being accepted as Scots or English, even if they have the same birthplace and accent as

---

[2] Tebbit also failed to notice that Scots and other non-English Brits were also often disinclined to support England and not just at cricket either. Sometimes Scots are extremely disinclined, supporting ABE ('anyone but England').

white people. True, birthplace is the most powerful, normally unchallengeable, criterion for national acceptance, but there remains a small, but significant, premium for being white. We saw that between the surveys of 2006 and 2008/2009, England had become slightly more 'rejectionist' of non-white claims to national identity, whereas in Scotland the views remained virtually unchanged. Our explanation for this was that a more explicit and continuing debate about 'who is a Scot?' north of the border has made it more difficult for the racialisation of politics to occur (the British National Party, for example, is historically much weaker in electoral terms in Scotland and, while less extreme, UKIP also has much less leverage).[3] We also found that the key factor in accounting for acceptance and rejection of claims was education; in other words, the more qualifications one has, the less likely one is to reject non-white claims to be 'one of us' in national identity terms.

Using the 2003 British and Scottish Social Attitudes Surveys, Asifa Hussain and Bill Miller reinforce this point (Hussain and Miller, 2006). They concluded that: 'Islamophobia is significantly lower in Scotland than in England – despite the growth of Scottish national identity and the advent of a separate Scottish parliament' (2006: 49). Further, they found a significant difference between 'exclusive Scots/English' (those who say they are Scottish/English and not British), and others:

Compared to those who identify equally with Britain and England/Scotland, the exclusively English are 27–30 per cent more likely to cite race ... as a necessary condition for being 'truly British' or 'truly English'. Yet in Scotland the exclusively Scottish are scarcely any more likely (only between 2 and 4 per cent) than those who identify equally with Britain and Scotland to cite race as a necessary condition for being 'truly British' or 'truly Scottish'. (Hussain and Miller, 2006: 62)[4]

In other words, 'exclusive' Scots, unlike their English counterparts, do not display high levels of Islamophobia. The authors conclude: 'Islamophobia is not only significantly greater in England than in Scotland, it is also much more closely tied to English nationalism

---

[3] At the Euro-elections in May 2014, UKIP got 10 per cent of the vote in Scotland, compared with almost 30 per cent in England.
[4] Their findings are broadly comparable to our own, albeit our focus was on 'non-whites' rather than specifically on Muslims (see Chapter 5).

within England than Scottish nationalism within Scotland – with which it hardly correlates at all' (Hussain and Miller, 2006: 65).

They claim, however, that Scotland does not come out of this with an entirely clean bill of health, for there is also what they call 'Anglophobia' to contend with. Hussain and Miller present summary data showing that Anglophobia exists in Scotland, albeit at a fairly low level, even lower than Islamophobia (see Hussain and Miller, 2006: table 3.1, p. 56).

This leads them to conclude that Scots in general, and what they call street-level Scottish nationalists in particular, have *different* phobias from the English rather than lesser ones, though in our opinion this is to exaggerate somewhat. The authors sum up their findings using an interesting metaphor:

For English immigrants culture is the bridge and identity the wall. Their culture is close to that of the majority Scots and more important, it is flexible ... but English migrants can't easily or quickly adopt a 'Scottish' identity.

For ethnic Pakistanis, culture is the wall and identity is the bridge. They identify quickly, easily, and in large numbers with Scotland; but they want to change Scotland by adding to the variety of Scottish culture and traditions. (2006: 169)

Are Scots ethnic or civic?

Much is made of the distinction between ethnic and civic in the literature on nationalism and national identity (see, for example, Yack, 2012), but as Jonathan Hearn has pointed out, 'to the extent that we understand "ethnic" as meaning "cultural" (as opposed to biological or based on some symbolic extension of kinship), all nationalisms, even the most civic and liberal, are ethnic' (Hearn, 2000: 12). He argues that such a distinction between ethnic and civic is far more about styles of argument about what nations are, and how social values are created, rather than defining actual types of nations or societies (2000: 194).

We can extend our comparative understanding of England and Scotland with reference to a different question in the 2003 survey data. The surveys in both countries asked for responses to the statement that: 'England/Scotland would begin to lose its identity if more Muslims came to live in England/Scotland.' We can use this question to get some purchase on the 'ethnic' or 'civic' aspects of national identities in the two countries. Focusing on respondents born and living in the respective countries (whom we refer to as 'natives'), we can confirm Hussain and Miller's finding that 'Islamophobic' attitudes are somewhat stronger in England than in Scotland.

There are clear differences between the English and the Scots, such that a majority of English natives (55 per cent) adopt what one might term the 'ethnic' viewpoint, agreeing or strongly agreeing with the question, compared with 41 per cent in Scotland. The respective 'liberal' or civic responses, based on those disagreeing with the proposition that Muslim in-migration would lead to a loss in national identity, are 25 per cent in England, and 39 per cent in Scotland. So whereas Scots split almost equally between ethnic and civic, the equivalent English split is roughly 2 to 1.

In the 2003 Scottish Social Attitudes Survey, there was a similarly worded question about English in-migration.[5] This time, even fewer, 33 per cent, of Scottish natives adopted the 'ethnic' response, compared with 46 per cent adopting the 'civic' one. If we combine this with the former question about the consequences of more *Muslims* coming to live in Scotland we can identify those who take a clear 'ethnic' view of the impact on Scottish identity by agreeing or strongly agreeing with both statements, and those who adopt a more liberal or civic viewpoint. We find strong associations between the answers people give. Thus, 29 per cent of Scottish natives have views which might be labelled 'ethnic' on both counts and 36 per cent 'civic'.[6]

So who are the 'ethnics' and the 'civics' among the Scots, and what accounts for their standpoints? 'Ethnics' tend to be older, to be lower class, to have no educational qualifications, and slightly more likely to say they are 'Scottish not British'. 'Civics', on the other hand, tend to be younger, of higher social class, and better educated, though there is little difference with regard to national identity. If we model these data, we find that above all, it is level of education, rather than age, sex, social

[5] The question asked for responses to the statement: 'Scotland would begin to lose its identity if more English people came to live in Scotland.' Given the relative sizes of England and Scotland, there was deemed by those running the British Social Attitudes Survey to be little point asking the 'Scottish' variant of the question in England.
[6] The next largest categories are 13 per cent who are neutral on both, neither agreeing nor disagreeing with the propositions, and 8 per cent who take an 'ethnic' view on Muslims but a 'civic' view on the English.

class or national identity, that differentiates the 'ethnic' from the 'civic' group.[7]

When asked who respondents thought should be entitled to a Scottish passport, only those considered to be 'truly Scottish', or 'anyone permanently living in Scotland' were the main categories of response,[8] and there is a strong association with the ethnic/civic divide. Thus, 73 per cent of those adopting a 'civic' standpoint said 'anyone permanently living in Scotland' should be entitled to a Scottish passport, and only 18 per cent of them 'only those truly Scottish'. The 'ethnic' respondents split evenly between 46 per cent and 43 per cent respectively, showing that even those who took a more 'ethnic' view were prepared to give marginal precedence to the 'civic' position – residence.

We might sum up our findings at this point by saying that, despite the official position of Scotland's elites and political classes that the 'civic' definition of who is a Scot is (or should be) the normative one of residence, a minority, but a significant minority, just under a third of Scots favour a more 'ethnic' viewpoint about who counts as 'one of us'.

Commonalities

We can now ask: with whom do people in Scotland and England compare themselves, and does it matter? We think that it does. If the Scots and the English felt they had little in common, then it would seem quite likely that ultimately the British state would not hold. On the other hand, if each feel they have a lot in common with the other, then although their primary allegiance is as we have seen to fellow nationals (Scots or English) rather than to fellow British citizens, the state would seem to be on firmer ground.

It has long been one point of view that the ties that bind people together in this island are those of social class with the result that party politics have historically been structured around the politics of class rather than nation. Unlike many continental European countries, 'British' politics, at least in England, Scotland and Wales, has lacked

[7] Using binary regression models on both ethnic and civic attitudes as we have measured them, we find that differences of age, sex, social class and national identity are not significant, but that levels of education are in both the 'ethnic' model ($p = 0.028$) and the 'civic' model ($p < 0.001$).
[8] Those who opted to say 'both', or 'it depends', or gave no answer numbered around 10 per cent of both 'ethnics and 'civics'.

significant religious cleavages, delivering a 'purer' form of class politics than elsewhere (Keating and McCrone, 2013: ch. 3). As Colin Crouch has observed, although social democracy is an international aspiration, it is deeply rooted in the 'nation-state' (Crouch, 2013: 164). Writers and politicians on the Left have been at pains to stress the significance of class and importance of class commonalities rather than national differences. That is, for example, the argument in Eric Hobsbawm's riposte to Tom Nairn's *The Break-Up of Britain* (Hobsbawm, 1977). Somewhat ironically, Nairn himself espoused the 'classist' position in his earlier work, as in this comment in an essay, 'The Three Dreams of Scottish Nationalism':

The Edinburgh baillie [magistrate] and the shipyard worker can both be joined in praise of Nationalism; but the nation and its culture belong to the former, not to the latter, and the triumph of a merely populist nationalism will signify a greatly strengthened grip of the real ruling class. (Nairn, 1988: 39)

The former Labour Prime Minister Gordon Brown made it one of his central arguments (see, for example, his speech to the Scottish Labour Party in September 2006). More dramatically and with familiar rhetoric, elements of the far Left have nailed their colours to the mast in the context of the 2014 Scottish referendum on independence, thus:

The prospect of a break-up of the UK and the division of the British working class on nationalist lines would be highly counterproductive in the struggle for socialism. We put forward the case for the unity of the working class in Britain and internationally, in the fight for the socialist transformation of society. (*Socialist Appeal: The Marxist Voice of Labour and Youth – Journal of the International Marxist Tendency* (IMT): www.socialist.net/a-marxist-view-on-scottish-independence.htm; accessed 18 September 2013)

The Labour Party's hostility to Scottish independence is to some extent driven by a belief that, for example, the Glasgow working class has more in common with that of Liverpool than it does with the Scottish bourgeoisie.

Does the evidence support these arguments that class identity outweighs national identity in Scotland, and what is the situation in England, where arguably national identity is more opaque? Interest in the 'class versus nation' question has generated a sequence of survey questions in Scotland since 1979, the date of the first, unsuccessful,

Table 7.1 *'Who do you feel you have most in common with?'*

	Scottish natives* (%)	English natives* (%)
Same class, different nation	32	43
Same nation, different class	68	57

* 'Natives' are respondents born and living in the same country.
Sources: Scottish Social Attitudes Survey, 2006; British Social Attitudes Survey, 2006.

devolution referendum on a Scottish 'Assembly'.[9] People have repeatedly been asked: 'whom do you identify with most: an English person of the same class, or a Scottish person of a different class?' In 1979, marginally more Scots put class identity before national identity (see McCrone, 2001: table 7.9, p. 167).[10] However, when the question was repeated in 1992, 1997 and 1999, a considerable majority opted for 'different class, same nation'. This further confirms what we saw in Chapter 3; there was a strengthening of national identification in the 1980s such that Scots were giving priority to being Scottish over being British.

We repeated the question in the 2006 surveys in both Scotland and England. In Table 7.1, we focus on the relative distributions of those who opted for these responses, excluding those who did not give a preference (17 per cent in Scotland, and 25 per cent in England).

The preference for 'nation' over 'class' is clear among Scots by a ratio of 2 to 1, but the English too share this view by a ratio of 1.3 to 1. Nor do these results reflect social class differences in the two societies because among those identifying themselves as 'working class' in Scotland and in England, the ratios are to all intents and purposes the same as in the wider population.[11] It appears that the propensity of both Scots and English 'natives' is to prioritise nation over class, albeit to different degrees. We accept that juxtaposing 'nation' and 'class' is in practice an artificial distinction, but it is a useful heuristic one in this context. We ourselves have observed that nation and class are, in a Scottish context, often proxies for each other (McCrone,

[9] Although a majority, 51.6 per cent, voted in favour, it was well short of the threshold stipulated by the Westminster government that 40 per cent of those on the electoral register had to vote in favour for the result to be accepted (the '40 per cent rule').
[10] By 44 per cent to 38 per cent.   [11] 2:1 in Scotland and 1.4:1 in England.

2001: 166–7), and in more recent articles Robin Mann (2012) and Arthur Aughey (2012) have made a similar point about England.

We have reached the point where we can address more directly the general issues of national identification and the political-constitutional issues which may arise. Who do 'nationals' think they have most in common with, and who the least? How do the Scots and the English view each other? Do the Scots think they have most in common with the English, and vice versa?[12]

We asked: '*Which of these groups do you, as a Scottish/English person, feel you have most in common with?*'[13]

'*And which do you have least in common with?*'

Respondents were offered a list of six options: English/Scots,[14] Welsh, Irish, American and French, as well as 'none of these' (Tables 7.2 and 7.3).

A number of features emerge. There is some commonality between the Scots and the English (28 per cent and 30 per cent respectively; Table 7.2), and despite claims that there is considerable antipathy between Scots and the English, it is striking that only 1 in 10 Scots say they have *least* in common with the English, and reciprocal feelings among the English are virtually non-existent. Overall the groups with which the two nations have least in common are remarkably similar. There are, however, some major differences as regards commonalities most strikingly with the Irish. Over half of the Scots say they have most in common with the Irish compared with only one-sixth of the English.

What conclusions can we draw from these data? Much depends on the propensity to see the glass as half-full or half-empty. On the one hand, there is little evidence of a lack of commonality between the two national peoples (although some indifference towards the Welsh), and a modest tendency for the Scots and the English to choose each other, albeit less than one-third of either Scots or English do so.

---

[12] The set of questions were asked in the British Social Attitudes Survey in 2008 (on which the English data are based) and the Scottish Social Attitudes Survey in 2009.

[13] These questions were asked in Scotland and England of everyone in the sample apart from those who said they were 'British not Scottish/English'. We included 'Americans' as people who shared the common language, and 'French' who did not but were near-neighbours.

[14] Scots were asked about the English, and the English were asked about the Scots.

Table 7.2 *'Which of these groups do you, as a Scottish/English person, feel you have most in common with?'*

	Scots (%)	English (%)	Scots–English (%)
American people	3	15	−12
Welsh people	8	19	−11
French people	3	2	+1
English/Scottish people	28	30	−2
Irish people	51	16	+35
None of these	8	16	−8
Base	1156	2080	

Table 7.3 *'And which do you have least in common with?'*

	Scots (%)	English (%)	Scots–English (%)
American people	27	27	0
Welsh people	7	5	+2
French people	50	56	−6
English/Scottish people	10	1	+9
Irish people	5	5	0
None of these	4	6	−2
Base	1149	2110	

Do these responses vary by social characteristics such as sex, age, social class and education? For the Scots, national identity plays a most significant part (Table 7.4).

There are strikingly clear differences. The more 'British' you say you are, the more likely you are to feel you have most in common with English people and the more 'Scottish' you are, the more likely you are to feel you have most in common with the Irish. To be sure, we do not know whether 'Scots' are treating 'Irish' as an undifferentiated category, making no distinction between the 'northern' and 'southern' Irish, who have quite different historical and religious connections with Scotland. This could then be linked to the religious persuasion of the respondent, and we do indeed find that Catholics are more likely than the rest of the sample to say they have most in common with the Irish (76 per cent compared with 51 per cent). However, those classified as of no religion, Church of Scotland and 'other Christian' also show significantly high associations; respectively, 49 per cent, 46 per

Table 7.4 *'Who do you have most in common with?', by national identity of respondent*

	% within each national identity category		
	Most in common with the English	Most in common with the Irish	Base
Scottish not British	18	61	379
Scottish more than British	24	56	414
Equally Scottish and British	40	38	328
British more than Scottish	69	26	35
All	28	51	1156

cent and 51 per cent say they have most in common with Irish people. It seems that having most in common with the Irish is strongly associated with feeling Scottish, and one might speculate that this is because Scots have more in common culturally with the Irish than they do with the English. Being a (Scottish) Catholic strengthens the association.[15]

We have just seen that the more 'British' you say you are, the more you are likely to say you have most in common with English people. Intuitively, one might also expect some connection with favouring Scottish independence but the effect is not a strong one and does not hold up in a binary regression model. This suggests that what drives the 'British' Scots to feel they have most in common with English people is not simply support for the Union.

And what of the English? As we saw, English natives feel they have most in common with the Scots (30 per cent), followed by the Welsh (19 per cent), the Irish (16 per cent) and Americans (15 per cent). You are more likely to say you have most in common with Scots if you are older and if you think of yourself as 'British'. There is a similar propensity among English people as regards the Welsh – the more British you are, and the more education you have, the more likely you are to think you have most in common with the Welsh.

---

[15] The 'religious effect' holds up in a full regression model independently of national identity, and both are significant at the 0.001 level.

The notional other – an overview

In this chapter, we have shown that national identity, like other social identities, is constructed in terms of the 'vis-à-vis'. Put simply, we know who we are by a process of comparing ourselves to other groups. We look for differences telling us who we are not; differences which may exist or which, regardless of objective evidence, we believe exist. We have focused, in the main, on Scottish–English identity relations, and discovered that the English look to the Scots as those with whom they have most in common, while for the Scots it is the Irish, although there is some reciprocity between the two largest national groups on the British mainland. We also saw that when people talked about national identity, some were prepared to use 'ethnic' language to make the point of difference: 'they are Celts; we are Saxons', 'it's genetic'; 'something to do with the blood system'; 'we are not them', and so on. These respondents were seeking a language and vocabulary to express what they saw as basic differences. As Eriksen puts it: 'ethnicity emerges and is made relevant through social situations and encounters, and through people's ways of coping with the demands and challenges of life' (1993: 1).

It would be quite wrong to assume that such purported differences are fixed. There are good examples of situations in which social relations between people change and become 'ethnicised'. For example, the social anthropologist F. G. Bailey (1996) has commented on the 'civility of indifference' which resulted in an 'amiable mix' of Serbs, Croats, Muslims and Slovenes in a small Croatian town which he studied in the early 1980s, only for that 'civility' to be violently overthrown in the 1990s in the context of the break-up of Yugoslavia. Similarly, in his 1950s study of Bisipara in India, he speaks of 'calculated restraint' between different ethnic and religious groups which had encouraged social relations to be pragmatic, quotidian and calculating, but which, by the 1990s, broke down into open conflict under the instigation of political entrepreneurs with axes to grind. Long-standing differences of caste were challenged in post-partition Bisipara by the implications of the 1949 Temple Entry Act which gave 'untouchables' access to all Hindu temples, presenting a potential flashpoint for caste conflict. Bailey observes:

The cultural performances that I saw – the political theater of a struggle for power between untouchables and clean castes – were carefully staged and insulated so that there would be no damaging fallout on a style of life so internalized in village habits that it needed no discursive articulation. (Bailey, 1996: 13)

These examples illustrate the fundamental point that such 'differences' are not inherent, nor necessarily conflictual, but they do provide the raw materials for conflict, be that purely political or violent, should conditions occur in which there is struggle for resources, be they material or cultural. As Peter Worsley points out: 'cultural traits are not absolutes or simply intellectual categories, but are invoked to provide identities which legitimise claims to rights. They are strategies or weapons in competition over scarce goods' (Worsley, 1984: 249).

Lying at the heart of so much discussion about 'national identity' we find the demos/ethnos trope. This distinction, sometimes expressed as 'civic/ethnic', has dominated the study of nations and nationalism at least since Hans Kohn (1944) gave it prominence in the 1940s. However, as we pointed out earlier, Jonathan Hearn has perceptively commented that these tropes do not classify real societies or nations, but are, in an anthropological sense, social and cultural myths, a point echoed by the political theorist Bernard Yack:

The myth of the ethnic nation suggests that you have no choice at all in the making of your national identity: you are what you have inherited from previous generations and nothing else. The myth of the civic nation, in contrast, suggests that your national identity is nothing but your choice: you are the political principles you share with other like-minded individuals. Real nations, in contrast, combine choice and cultural heritage. (Yack, 2012: 30)

The myth of the civic nation is thus the myth of *consent*; the myth of the ethnic nation is the myth of *descent*. Employing ethnic/civic tropes is to use a powerful figure of speech, and such 'myths' are signposts, and even help us to analyse the social and demographic beliefs and differences of those who use them. Yet the last sentence of the above quotation goes to the heart of the matter; individuals make choices with regard to national identity and express them in different ways. The language of identification and association is complex and nuanced. Thus, and it bears repeating, Hussain and Miller, in their study of Muslim and English migrants to Scotland, comment that for Muslims, national identity is a 'bridge' and culture is a 'wall', whereas for English migrants it is the other way round. Thus, Muslims can claim to 'be Scottish' (the bridge) whereas the English may share symbolic repertoires of culture, but cannot think of themselves (or, for that matter, be accepted as) 'Scottish'.

Our empirical focus on the 'notional other' in this chapter shows that much is implicit, and further, is a complex mix of various forms of social identification including ethnicity, nationality and social class. If who we are depends on whom we compare ourselves to, and in what contexts, then similarities may be just as important as imputed differences. Put simply, we may consider ourselves to be 'like' a notional other for some purposes and in some contexts, and 'not like' them in others. The search for difference may sometimes mask the search for similarity. 'They' may be 'us' for some purposes and 'them' for others. In other words, we may be faced with a 'wandering we' in national deixis (Petersoo, 2007).

For the English and the Scots, there is a possible shared identity on offer: be British. Given the political developments in these islands, this way of avoiding potentially stressful relations with 'the other' may be made more difficult by a purported crisis of Britishness, a question mark over the issue of how 'British' people feel. There is an argument that one aspect of the constitutional crisis in these islands relates to the alternative national identities the people of the United Kingdom can adopt. In short, do 'British' citizens feel British any more, and if not, what is the prognosis for the state they share? This is the puzzle we will tackle in the next chapter.

8 | *A manner of speaking: the end of being British?*

Running through many of the chapters, but not discussed in detail, has been a key question: have the British ceased to be British? In particular, have the Scots ceased to be British, and are there signs that the English are following the same path away from state identity?

Much of the debate about 'being British' has been driven by the politics surrounding the constitutional future of the United Kingdom and the possibility of Scottish independence. This has led, ineluctably, to assertions about the declining impact of Britishness, and how, in the interests of the Union, it might be revived. Being British has largely been viewed through a political prism, as if this is its essence, and that it follows that anyone claiming British identity must be in some sense a 'unionist', and not a 'nationalist'. This, we believe, is too simple a view, or, rather, it ought to be seen as a question for research rather than being taken for granted.

It would be highly unusual, possibly unique, if a large and established West European state in the twenty-first century lost significant territory through secession. If Scotland did leave the Union following the referendum vote in September 2014, or in the following few years, there would still remain a large state of around 55 million people. Such loss of territory would be especially unusual if secession were to be driven by shifts in national identity, although we shall argue later that this intuitively attractive hypothesis is improbable.

The Union Flag is a recognised global icon, flown and worn on a variety of occasions. And yet, like the owl of Minerva, it may fly only at dusk, presaging the end of being British. Given that 'Great Britain', in one form or another, has existed since the Union of the English and Scottish Parliaments in 1707, it comes as something of a surprise

A shorter version of this chapter was published as 'The end of being British?', in *Scottish Affairs*, 23(3) (Bechhofer and McCrone, 2014b). We are grateful to the publishers, Edinburgh University Press, for agreement to publish.

to learn that it was not until 1975 that a historian, J. G. A. Pocock, a New Zealander to boot, made 'a plea for a new subject' – *British* history (Pocock, 1975). To be sure, there had been much English, Scottish, Irish and Welsh history, claimed Pocock, but nothing that could correctly be called *British* history. As non-historians, this is not our quarrel. It is, however, in a different way also a question for social scientists. The ease with which writers equated 'society' with 'state', or unthinkingly described Britain as a 'nation-state' seems to have largely passed.[1] The shift in language and nomenclature indicates a certain unease with assuming that 'Britain' rules the waves, at least those around these shores. The term 'British Isles' no longer is meaningful, for it excludes most of the island of Ireland these days – and 'these isles' seems too vague – although it is a fair bet that Pocock's prefer-ence for the 'Atlantic Archipelago' is unlikely to catch on as a substi-tute. Strictly speaking, 'Great Britain' and the 'United Kingdom' are of course not coterminous, for the former excludes Northern Ireland. This is a mite pedantic, however, especially as the majority 'commu-nity' in Northern Ireland proclaims its Britishness with some vehe-mence, and ironically, much more than those living on the 'mainland'.

If the relation between Britishness and attitudes to constitutional change is a topic for research rather than assertion, the starting point must be how people in these islands describe themselves specific-ally with regard to Britishness. The traditional 'forced choice' ques-tion about national identity offers a choice between 'British' and the appropriate national identity, asking 'and if you had to choose, which one **best** describes the way you think of yourself?' In 2007,[2] in terms of 'best' descriptor, the English split 47 per cent 'national' and 38 per cent 'British'; the Welsh 56 per cent and 31 per cent; the Scots 77 per cent and 14 per cent, with a majority opting to be British only in Northern Ireland (56 per cent to 33 per cent Irish). That last fig-ure is somewhat misleading because the people of Northern Ireland split neatly along 'community' lines such that Protestants are 75 per cent 'mainly British', while Catholics are 75 per cent 'mainly Irish', so the 'majority' option in the Province for being British is somewhat

[1] Using the term 'nation-state' for the UK may have passed, but assuming the cultural-political homogeneity of the British state arguably has not.
[2] 2007 is the latest year for which data on all four UK territories are available at a single time-point.

artificial. The point remains, however, that on these figures, neither the English, nor the Welsh nor the Scots give priority to being British. This would seem to support the contention that the British are, in the main, no longer 'British'. And yet all is not what it seems. For example, the 'free choice' question which precedes the forced choice one[3] shows that two-thirds of the English (67 per cent), over half of the Welsh (58 per cent) and almost half of Scots (43 per cent) are prepared to say they are British *as well as* 'national'. In other words, 'forced choice' may have the merit of clarity but reveals none of the complexity which attaches to national identities.

The politics of British identity

What has given credence to the idea of a 'crisis' of Britishness is politics, primarily Westminster politics. The view that being British is in some difficulty and needs boosting is shared by politicians from the two main British political parties. Here is an extract from David Cameron's first speech to the Conservative conference as prime minister: 'When I say I am Prime Minister of the United Kingdom, I really mean it. England, Scotland, Wales, Northern Ireland – we are weaker apart, we are stronger together, and together is the way we must remain' (6 October 2010). He was following in the footsteps of his predecessor Gordon Brown, who had made a series of speeches in this century (8 July 2004; 13 December 2005; 9 September 2006; 13 January 2007), and published an edited book (with Matthew d'Ancona) at the end of the decade entitled *Being British: The Search for the Values that Bind the Nation* (2010). In a speech supporting the campaigning group 'United Kingdom with Labour' in Glasgow on 2 September 2013, he said:

Like the Clydesiders, we have a big idea – we believe that Scotland is a nation – no-one is more proud of being a Scot than I am and I suspect you are ... But I believe also that we are part of something bigger. That just as the Clydesiders proposed that for pensioners and for worker's rights and for the unemployed and for those people in need of health care – we should pool and share the resources of all the nations of the United Kingdom, I believe that we should set down that very clearly as the purpose of the

---

[3] The question asks: 'Please say which, if any, describes the way you think of yourself. Please choose as many or as few as apply: British, English, European, Irish, Northern Irish, Scottish, Ulster, Welsh, other answer.'

Union. (www.scottishlabour.org.uk/blog/entry/a-positive-principled-and-forward-looking-case-for-the-union)

Brown is a politician who has talked frequently and at length about what it means to be British.[4] His attempts to distil a particular set of 'British' values were attacked on the grounds that democracy and the rule of law on which he focused were not unique to Britain. Critics pointed out that Brown's attempt to forge a British Way owed too much to specifically *English* rather than British experiences: Magna Carta, the fourteenth century Peasants' Revolt (Wat Tyler), the 1689 Bill of Rights all predated the 1707 Treaty of Union, which created Great Britain (Lee, 2007). As a politician with a doctorate in history, Brown's response would surely have been that those items were woven into the fabric of being British, and were complemented by specifically non-English experiences. No matter. Brown's problem was that of any non-English prime minister, namely, to distance himself from his Scottish credentials, especially when Scottish nationalism was in the ascendant. There was also a back story in the claim that the then-Labour government was dominated by a 'Scottish Raj', a Cabinet containing a disproportionate number of Scottish politicians.[5] This came at a time when the so-called West Lothian Question was becoming more salient, raising the issue that Scottish MPs at Westminster could vote on English laws, but they, and more particularly English MPs, could not influence those which had become the responsibility of the Scottish Parliament in Edinburgh. This was, for critics, something of a perfect storm: a Scottish prime minister, a Scottish-dominated Cabinet, a Scottish Parliament making laws for Scotland, the West Lothian Question and a claim from the Conservative opposition that Labour had a weak mandate in England. In making that claim, they chose to ignore the fact that Labour won 90 more seats than the Conservatives in England at the 2005 British general election, albeit Labour's share of the popular vote was 0.2 per cent lower.

Some critics went further, claiming there was something of a conspiracy against the English. In the words of one political scientist,

---

[4] During the referendum campaign leading up to the 2014 referendum on Scottish independence, Brown published a further book called *My Scotland, Our Britain: A Future Worth Sharing* (2014).

[5] www.telegraph.co.uk/news/uknews/1485591/Britain-run-by-Scottish-Raj-claims-Paxman.html, accessed 9 July 2014.

Brown's British Way was 'a means of negating both England as a political community and Englishness as a source of political identity' (Lee, 2011: 161). The British Way was not simply a rhetorical ploy to let a Scottish prime minister off a difficult political hook, but a tactic with substance; a deliberate device to keep England in its place. Thus, Lee argues that the British Way had two purposes: the first that

> by constantly reminding the English of the importance of Britain and Britishness, as the focal point for their own patriotic purpose, it would perpetuate the mindset that had long conflated and confused the identity and interests of England with those of the United Kingdom, and Englishness with Britishness.

And the second that

> by subordinating England and Englishness to Britain and Britishness, the British Way might divert the attention of English voters away from the fact that they had been bypassed by New Labour's most important constitutional reform, and thereby denied the parallel extension of democratic citizenship and accountability which had been extended to the people of Wales, Scotland and Northern Ireland. (Lee, 2011: 160)

To put it more bluntly, the (con)fusion of England and Britain was taken to be a deliberate political strategy to obfuscate identity issues; and second, the aim of this policy was to prevent 'home rule' for England. Writing in the same volume, Lee's fellow political scientist, Colin Copus, goes further: 'trying to ignore England, or worse still, to regionalise ... is anti-democratic and further evidence – were it required – that the success of Britain, as a multinational state, has always relied on the sacrificing of England' (2011: 209). It is not clear how 'regionalising' England is 'anti-democratic', especially in the light of claims that England is too large and diverse to be governed from Westminster. There seems much wishful thinking here that, as a result of these slights, in Copus's words, 'the slumbering giant that is England is beginning to awake' (2011: 210); distinct echoes of Chesterton's 'we are the people of England that never have spoken yet'.

Such claims are a long way from Anthony Barnett's contention which we outlined and discussed in Chapter 3 that the English are baffled when they are asked to think of 'English' and 'British' as separate identities (Barnett, 1997 and 2013). In contrast, and on the basis of very little evidence, Lee and Copus are saying there is a deliberate policy of depressing (suppressing?) English political identity in favour of being British because the (British) state could not survive such an expression of *national* identity were it to gain strength and momentum. To be sure, there are more sophisticated versions of such an account, notably that of the political theorist Bernard Crick who argued that 'From political necessity English politicians tried to develop a United Kingdom nationalism and, at least explicitly and officially, to identify themselves with it, wholeheartedly' (1989: 29).

We might take the more pragmatic view that the emergence of the 'English Question' took place under the peculiar and highly specific circumstances of a Labour government at Westminster, with a Scottish prime minister and a 'Scottish' Cabinet, in the context of an ascendant Scottish National Party north of the border, together with Scottish (and Welsh and Northern Irish) devolution, under attack from an overwhelmingly, in terms of seats, 'English' Conservative Party. It is going well beyond the evidence to suggest that it was a matter of deliberate policy and that there is something of a conspiracy to prevent the rise of English political identity. As we saw in Chapter 6, two-thirds of English people are content to be governed by a *British* Parliament rather than a specifically *English* one. What we may be seeing from writers such as Lee and Copus is an attempt to formulate a proto-national English agenda to undermine Britishness at a time when the future of the British state is in question as Scotland has a referendum on independence in 2014. In any case, it is a good example of the salience of the politics of national identity in these islands.

An event such as the Scottish referendum inevitably raises the question of 'being British' and its political implications. For example, the political scientist Vernon Bogdanor has commented that

those choosing the separatist option in the 2014 referendum would be proclaiming that the two identities [Scottish and British] are incompatible, just as, when Ireland became independent in 1921, it signified that the identity of being Irish was incompatible with a British identity. (*The Guardian*, letters, 8 April 2013)

One reader replied:

In the referendum, Bogdanor writes, 'Scots will decide whether they want to remain British.' This is untrue. Most Scots identify as British and Scottish, and the referendum is not a choice between these identities. Just as an

Independent Norway is still Scandinavian, so an independent Scotland will still, in a cultural and geographical sense, be British. (E. Bulmer, 31 March 2013)[6]

Here we can see the nub of the debate in formal terms. For Bogdanor, being 'British' is equivalent to being a British citizen, and for him, by voting for independence, Scots will have given that up *de jure*. For his critic, being British is likely to survive a break-up of the state because it is not, in essence, simply a 'political' form of identity but, rather more, a cultural and historical one.

The UK government subsequently issued a position paper stating that

The UK has historically been tolerant of plural nationalities, and therefore it is likely that it would be possible for an individual to hold both British and Scottish citizenship. However, under current rules British citizens living outside the UK cannot pass their British nationality on more than one generation. So the children of British citizens living in an independent Scottish state would be British citizens, but their children and subsequent generations would not be. (www.gov.uk/government/uploads/system/uploads/attachment_data/file/274477/scotland_analysis_borders_citizenship.pdf, accessed 9 July 2014)

Coupled with the views of the current Scottish Government[7] that all British citizens habitually resident in Scotland, as well as Scottish-born British citizens currently living outwith Scotland, will also be considered Scottish citizens, it is likely that being Scottish and British would survive for some time after a vote for Scottish independence.

One can argue, then, that it is the *political* agenda, the politicisation of national identity which has driven this debate, rather than any major shifts in national identification among the 'British' themselves. Later in this chapter we will explore how 'British' the Scots (as well as the English) really are. Further, if in the event of Scottish independence some people continued to call themselves 'British', then as social scientists we ought to treat that as an interesting research question, rather

---

[6] Elliot Bulmer is described as research director of the Constitutional Commission. His full reply is available at www.theguardian.com/commentisfree/2013/mar/31/scotland-referendum-sovereignty-identity-bogdanor (accessed 9 July 2014).
[7] *Scotland's Future: Your Guide to an Independent Scotland*, Scottish Government, November 2013.

than dismissing it as a failure to comprehend constitutional law. It seems to us more of a sociological issue than a political one, although, based on our previous interview evidence, we might expect that some of those holding on to being 'British' would be making a 'unionist' statement by so doing. Little wonder, then, that issues of national identity have been caught up in matters of citizenship and politics. Given the amount of political attention focused on 'being British', we might even conclude that it is in something of a crisis, not only in Scotland, but also in England. In the next section, we focus on the contention that being British is in decline on both sides of the border.

How 'British' are the British?

If we use the classic Moreno question to examine this issue, we would conclude that being 'British' in some form is actually the norm, and only those who say they are 'Scottish not British' – about one-third of respondents in Scotland – deny being British. On the other hand, as we mentioned earlier in the chapter, a simple 'forced choice' question offering respondents only a dichotomy – are you Scottish/English/Welsh *or* British – tends to oversimplify identity choices in order to get a straightforward priority listing. Thus, if we look at the mean percentages of *forced* choices of English and Scottish people between 1996 and 2009, less than half of English people would choose 'British', and only 1 in 5 of Scots would do likewise. One could then easily come to the opposite conclusion and claim that the majority of the English and the Scots are not 'British' to any significant extent. There is yet another way of looking at this. If we focus on the 'free choice' results, where people can give more than one identity if they so choose, we then find that 'British' is a far more popular option. More than two-thirds of the English say they are British (virtually the same percentage say they are 'English'), whereas the Scottish figure is 50 per cent (over 80 per cent say they are 'Scottish'). We could now reasonably argue that it is not a question of *either* being 'national' or British, but being both. Depending on how one chooses to look at the issue, one can reach three very different conclusions. Those people who wish to use the surveys to make political points can, by choosing one set of data and ignoring the rest, argue the case for or against declining Britishness. Politicians, like social scientists, would be well advised to heed something our one-time colleague, the late Tony Coxon, was fond of saying: 'it depends whether you know what you want or whether you want to know'.

It is also clear that these simple labels do not tell us what being Scottish, English or British *means* to people. Further, more nuanced evidence that the Scots are not averse to 'being British' is to be found in the 2003 British and Scottish Social Attitudes Surveys. Respondents were asked: 'How proud are you of being British?' More people in England said 'very proud' (41 per cent) compared with 23 per cent in Scotland. Nearly 8 out of 10 in England expressed *some* pride in being British (79 per cent) compared with two-thirds in Scotland (64 per cent). Despite the differential, such data certainly do *not* suggest a strong rejection of being British in Scotland. We developed this line of argument further in the 2005 Scottish Social Attitudes Survey by offering respondents three statements with which they were asked to agree or disagree on a five-point scale (strongly agree, agree, neither agree nor disagree, disagree, strongly disagree):

- 'to be "British" is to be proud of Britain's past and the strong part it has played in shaping the world';
- 'to be "British" is something to be ashamed of because of the poor treatment of the people who lived in the former colonies of the British Empire';
- ' "British" is a label that unites all peoples living in Britain today regardless of colour, creed and nationality'.

For convenience, in Table 8.1 we abbreviate these as British Past; British Empire; and British Unity.

We find (Table 8.1) that well over half (56 per cent) said they were proud of Britain's past, not ashamed of Empire (57 percent), and took a positive view of 'British' as a unifying symbol (59 percent). Even among the 'Scottish not British' group, the equivalent figures do not show a dramatic fall: respectively, 49 per cent, 48 per cent and 52 per cent. Even among those people who, in identity terms, deny that they are British, and whom we might expect to be the most hostile, we find fairly positive attitudes towards Britishness.

This ambivalence is nicely expressed by comments from individuals interviewed in our nationals and migrants study:

Sometimes, again when you go back into history and there are things that we were once proud of, proud of the fact that there was once a British

Table 8.1 *Pride in Britain among Scots (% by row)*

	Strongly agree	Agree	Neither agree nor disagree	Disagree	Strongly disagree	Don't know
British Past	15	41	24	16	3	1
	56*			19**		
British Empire	2	17	22	45	12	2
	19*			57**		
British Unity	9	50	11	24	5	1
	59*			29**		

N = 1549. *Strongly agree + Agree. ** Disagree + Strongly disagree.
Source: Scottish Social Attitudes Survey, 2005.

Empire and that Britain had done so much, but then when you realise the cost and how much the people from the colonised lands had to suffer, I'm not so proud. In the time of the British Empire, they achieved a lot, they found a lot, they discovered a lot, but on the other hand there was a lot of things that they shouldn't be proud of.

However, this person goes on to say:

I'm not at all ashamed of being British, not at all. I suppose every country had a go at getting what it could for itself, at one time and making other people suffer. I'm not really ashamed but I'm very conscious of the fact that everything that Britain did is not something that we should be proud of.

Another respondent echoed these views:

Some of the things that happened we can't be proud of but I don't think any other country would have a clean sheet in this respect … Overall, I think the fact that the Commonwealth has replaced the Empire shows that there were some reasonable ties between the countries that were dominated, as it were. They never wanted to separate off entirely.

As we might expect, there is some variation by political party support. Conservative supporters in large part agree with the statement about the British past (74 per cent), and disagree (also 74 per cent) on the second item (British Empire). Yet, even SNP supporters show sizeable numbers supporting these 'British' options: 44 per cent and 52 per cent respectively, with Labour (61 per cent and 57 per cent) and Liberal Democrats (57 per cent and 62 per cent) somewhere in the middle. It is remarkable that just over half of SNP supporters say they are not ashamed of Empire, and although they are less enthusiastic about the British past than supporters of other parties, those Scottish nationalists taking a positive view outnumber the negatives by 1.5 to 1. As many as half of SNP supporters are prepared to accept 'British' as a label that 'unites people living in Britain today', although support-ers of other parties are unsurprisingly somewhat more likely to do so.[8] Taking these data together, the dominant impression is that a majority of people in Scotland and in England take some pride in being British and that in Scotland, although people are more strongly 'Scottish', they also do not take a negative view of Britain's past, its erstwhile Empire or in seeing 'British' as a multicultural and unifying label.

Symbols of Britain

It is, however, possible that the Scots and the English have different *conceptions* of 'British'. It could be that they were referring to differ-ent things when they spoke about being British. One way of exam-ining this is to look at one aspect of the content of Britishness, that is, symbols of British culture. Accordingly, in the British and Scottish Social Attitudes Surveys of 2006, we asked people to select from a list of such symbols which one they thought most important to British cul-ture, and which the second most important. The point of this exercise is worth underlining. We were asking respondents what they saw as important symbols of Britain, *not* how they related them to their own sense of being British, nor whether they approved of a particular sym-bol. Thus, someone may think of the British monarchy as an import-ant symbol of Britain, while personally holding republican views and disapproving of the institution. For 'British culture',[9] we included

---

[8] Labour supporters (68 per cent) are most inclined to agree, followed by Liberal Democrats (62 per cent) and Conservatives (57 per cent).

[9] We might be challenged that the British 'culture' implies culture narrowly defined as 'arts' (*kultur*). Had we used a word other than 'culture' (and there is no ready alternative), we might have found even more respondents opting for items of 'political culture' – democracy, monarchy, fair play. In another set of questions we asked as appropriate about 'English/Scottish culture'. Again, using the term 'culture' might have predisposed respondents to answer in narrowly

Table 8.2 *Symbols of British culture among English and Scots**

Symbol of British culture	English respondents	Scottish respondents	English–Scottish percentage difference
British democracy	61 (1)	61 (1)	0
British monarchy	45 (2)	39 (3)	+6
British sense of fair play	41 (3)	50 (2)	–9
British flag (the Union Jack)	27 (4)	20 (5)	+7
British sporting achievements	19 (5)	23 (4)	–4
British national anthem ('God Save the Queen')	11 (6)	8 (6)	+3
*Base***	2095	1131	

* The figures given first are the percentage of respondents mentioning the item as one of their two choices; those in brackets are the rankings within the column.
** Those who did not answer the question, or said 'don't know' or 'can't choose' (in England 9 per cent and in Scotland 13 per cent) have been excluded.

the following symbols: British sporting achievements, the British flag (Union Jack), British democracy, British monarchy, British sense of fair play, and the British national anthem ('God Save the Queen'). The key finding (Table 8.2) is that the English and the Scots broadly agree on the ranking of the symbols of British culture, and that a constitutional symbol, 'British democracy', is seen as the most important, and equally so in both countries.

The other constitutional symbol, the monarchy, which ranks second in England and third in Scotland, has a fairly high salience, even in Scotland.[10] The least chosen symbols are those of conventional cultural iconography (the anthem, sport and the flag). The English are slightly

'cultural' as opposed to 'social' or 'political' terms. Nevertheless, all the lists included items framed in a broader sense, and interviewers reported that *both* sets of questions worked in these terms. This reinforces our sense that the use of the word is justified.

[10] The Scottish National Party is sensitive to this issue, wisely, it seems, from these data. While there is a strong republican tendency in the party, the general expectation is that if Scotland became independent, the monarch would retain some symbolic position.

more likely than the Scots to choose the symbols of Union, the British Flag and the monarchy, while the Scots are more likely to value 'fair play' than the English, in the list of British symbols. While *statistically* significant, these differences can only be regarded as small in the light of the broad agreement between the English and the Scots shown by the data.

But, surely, one might think, asking respondents to identify British symbols is influenced by people's *own* sense of national identity, especially those with a strong sense of being English or Scottish. If national identity affected the results we would expect clear differences to appear in the 'Mainly English' and 'Mainly Scottish' categories. They do not, as we can see in Table 8.3.[11]

Looking first at *the rankings*, they are remarkably similar across all the columns. Both within each nation and across the nations the rankings differ very little.[12] The 'Mainly Scottish' respondents are considerably more likely to choose 'fair play' than the 'Mainly English'. The other differences between the 'Mainly English' and the 'Mainly Scottish' groups (a greater tendency for the English to choose the British Flag and a lesser tendency to choose' sport) are smaller. Across the board, then, it seems that the expected relationship does not hold; a person's sense of their *own* national identity makes little difference to their perceptions of the important symbols of British culture. Further, people do not have to think of themselves as 'British' to rate 'British' symbols in a broadly similar way to other persons living in Britain.[13]

But surely, someone might reply, the *decline* in British self-identification in recent years in both Scotland and England is well documented (Bechhofer and McCrone, 2008), and this probably has had an effect on the importance people attach to symbols of national culture. This is an example of people treating something as self-evident, usually without the benefit of evidence. There is a long and valuable tradition in sociology of showing that what is taken for granted often turns out not to be the case (Lazarsfeld, 1949; Bechhofer and McCrone, 2009a). It would seem 'obvious', for example, to think

---

[11] We are not able to compare directly responses by the 'Mainly British' groups because this category is very small in Scotland.

[12] *Within* both Scotland and England, none of the differences are significant at the 0.01 level.

[13] There are no statistically significant effects at all in England. In Scotland, in the case of 'British democracy' and 'British fair play' there are some differences which do not affect the overall conclusions. The detailed analysis can be found in Bechhofer and McCrone (2013).

Table 8.3 *Symbols of British culture by national identity**

Symbol of British culture	English respondents (% by column)			Scottish respondents (% by column)		
	Mainly English	Equally English and British	Mainly British	Mainly Scottish	Equally Scottish and British	Equally, mainly or only British combined
Democracy	60 (1)	55 (1)	67 (1)	60 (1)	66 (1)	65 (1)
Monarchy	44 (2)	48 (2)	37 (3)	38 (3)	39 (3)	40 (3)
Fair play	38 (3)	39 (3)	44 (2)	50 (2)	49 (2)	49 (2)
Flag (Union)	27 (4)	26 (4)	24 (4)	21 (5)	19 (4=)	20 (4)
Sport	19 (5)	19 (5)	17 (5)	24 (4)	19 (4=)	18 (5)
National anthem	12 (6)	11 (6)	10 (6)	8 (6)	9 (6)	8 (6)
*Base***	784	974	270	808	255	300

* The figures given first are the percentage of respondents mentioning the item as one of their two choices; those in brackets are the rankings within the column.

** Those who did not answer the question, or said 'don't know' or 'can't choose' have been excluded, as have all respondents not choosing a national identity.

that the English and the Scots would differ in the way they rank state and national symbols, and especially that national identity would be an important discriminator. British politicians, such as Gordon Brown and David Cameron, for example, might well believe that emphasising British symbols will halt or reverse what they see as the decline in 'Britishness' and in turn strengthen the British state. Our research, however, shows that this is misguided; the importance both the English and the Scots attach to symbols of British culture is not associated with their *own* self-identification. People's *own* national identity is, by and large, irrelevant when it comes to choosing which British symbols are important. We have shown that 'Britain' remains a salient and meaningful frame of reference, even though more and more people in England and Scotland do not define their own identity *primarily* as British. If, then, they choose not to define themselves as British, this is a positive decision, not because they think Britishness is a concept devoid of meaning. Being British *does* have content and meaning in terms of important symbols of British culture which are widely and normatively held. Furthermore, in *both* countries it takes an institutional and constitutional form: the institutions of democracy and a (constitutional) monarchy together with a sense of fair play, rather than a narrowly 'cultural' form, associated with orthodox icons of flag and anthem. In short, we cannot draw the conclusion that the relative decline in people's willingness to describe themselves as 'British' is the result of not knowing what the term means.

How Scottish are the Scots?

It is fair to say that the debate about Britishness has been largely driven in recent years by the presumption that the Scots in particular are ceasing to be British. And yet, whatever the views of those who are influenced more by ideology than evidence, we have seen that the 'end of Britishness' thesis is on shaky ground if one reviews the evidence across the board. It is, however, undeniable that while people in Scotland (as in England) *comprehend and think important* what is meant by British culture and its symbols, and most of them acknowledge some degree of Britishness in their identity, fewer choose to define themselves as British. The Moreno question, with which readers of this book will by now be very familiar, has been widely used in situations where constitutional change is an issue. It is helpful, then, to set national identity in context, to see whether and how the peoples of the United Kingdom compare with others. Unsurprisingly given its origins we have found examples of the five-point Moreno scale used in Scotland, England, Wales and Northern Ireland, as well as in Catalonia and Euskadi (the Basque Country) in Spain, Flanders and Wallonia in Belgium, and Quebec in Canada (see, for example, *Revue Internationale de Politique Comparée*, 2007). Trying as far as possible to standardise on survey year, we obtained the results shown in Table 8.4.

On the basis of these figures, we can see that, when compared with other nations, Scots are much more inclined to prioritise their 'national' over their 'state' identity. These data are taken from just one year but other data show that the ratio of 8:1 is not out of line with other years. Euskadi and Catalonia[14] come next with a ratio of 4:1, with Quebec and, interestingly, England next at 3:1.

The Moreno scale has served the research community well but its five-point scale is relatively unsubtle. Respondents who do not wish to choose the extreme points ('Scottish not British' or 'British not Scottish') are only offered one point between the two extremes and the mid-point ('Equally Scottish and British'). So we wondered what would happen if one stretched the scale from five to seven points, but left four of the seven points unmarked. Respondents are then offered a scale running from 1 to 7, on which 1 is marked as 'British not Scottish', 4, the mid-point, as 'Equally Scottish and British', and 7 as 'Scottish not British'. Points 2, 3, 5 and 6 are left unmarked. People who do not see themselves as 'Equally Scottish and British' or unequivocally British or Scottish can thus choose between a point which is closer to an extreme or to the mid-point. If we assume that both this seven-point scale and the usual Moreno scale can be regarded as interval scales, and correlate the two, there is a considerable, if far from perfect, association.[15] We can, then, be reassured that the long established Moreno (five-point) scale, while obviously somewhat cruder than the seven-point scale, continues to serve its purpose well, and that its use as a survey measure of national identity is justified.

[14] In recent years, the ratio in Catalonia has risen to 8:1 (see, for example, Baròmetre d'Opinió Política, Centre d'Estudis d'Opinió, 29, 1a onada 2013). This reflects heightened tension between the Catalan and the Spanish governments (see Marti, 2013).
[15] The Pearson correlation coefficient is 0.646 and statistically highly significant.

Table 8.4 *National identity in comparative context (% by column)*

	Scotland 2006	England 2006	Wales 2007	Northern Ireland 2007	Catalonia 2010	Euskadi 2005	Flanders 2004	Wallonia 2004	Quebec 2007
Only national	35	22	24	19	19	26	7	3	19
National > state	32	17	20	17	28	22	29	11	32
National = state	22	47	32	17	39	35	45	31	28
State > national	4	8	9	24	5	7	8	13	12
Only state	4	6	9	19	7	4	11	42	7
National:state*	8:1	3:1	2.5:1	1:1.2	4:1	4:1	2:1	1:4	3:1
Base	1456	2431	884	1160	2000	1495	517	310	1251

* Ratio of 'mainly national' (rows 1 + 2) to 'mainly state' (rows 4 + 5).

The Moreno question does not, however, enable researchers to say anything about *how strongly* people feel about their declared national identity. Feeling strongly about Scottish identity and giving it *preference* does not preclude also feeling strongly about British identity. One person might, for example, feel strongly about both, and another weakly about both, and they might give similar answers, yet we would not know whether it really mattered to each of them. In the next section we examine this issue more closely using data never previously collected.[16]

How strongly British are the Scots?

The implicit purpose of the conventional Moreno five-point scale is to get people to compare how Scottish and/or British they feel. That leads us to assume that very few Scots, possibly only 1 in 10 or less, feel British (Table 8.4). What if that is simply a reflection of the way the question is asked? To test this, we asked respondents to express the strength of their 'national' and their 'state' identity on two *separate* scales without thinking about how the two identities related to each other.[17] Both scales had seven points, and ran from (1) 'Not at all Scottish' ('British' on the other scale) to (7) 'Very strongly Scottish' ('British' on the other scale). If we treat the scales as interval, the strength of Scottish identity is heavily skewed towards the strong end of the scale (mean of 6.12 and standard deviation of 1.11). The British scale is much more evenly distributed (mean of 4.27 and standard deviation of 1.78). Scots, perhaps unsurprisingly, feel more strongly Scottish and less strongly British. The next question then is whether feeling strongly Scottish must mean feeling weakly British. We can

[16] The results reported here are from the 2011 Scottish Social Attitudes Survey. Where appropriate, we have compared them with the 2012 survey results in order to be reasonably confident about the reliability of what is a relatively untested technique for getting at national identity. Unfortunately, these questions have not been asked in a British Social Attitudes Survey, so we do not know how the English would respond. It seems to us likely that the implications of the strength (or weakness) of 'national' and 'state' identities would not be very different.

[17] It is important to remember in what follows that respondents answered the relevant questions at different points in the questionnaire. We deliberately ensured that the context was different. Had the questions been asked consecutively it is more likely that respondents would have related the two questions to each other.

begin to examine that question by cross-tabulating the Moreno scale with the two scales measuring strength of feeling.

Even among those respondents claiming to be 'Mainly British', only about 1 in 10 place themselves below the mid-point, point 4, of the strength of *Scottish* identity scale. We can see then that 'Brits' in Scotland are not claiming this identity at the expense of saying they are Scottish. Further, of those who claim to be 'Equally Scottish and British', over half place themselves on point 6 or 7 of the 'Scottish' scale, and this rises to almost 9 out of 10 of the 'Scottish not British' group. Given that the mean value on this scale is 6.12, we might have expected this weak association between national identity and strength of Scottishness. It is, however, when we ask how strongly *British* these groups feel that the value of these new scales becomes apparent.

A claim to be Scottish *not* British does not necessarily imply a weak British identity as strict logic would lead one to expect, for nearly half (46 per cent) score at the mid-point or more towards the strongly British end of the scale. There is no neat correspondence between where people put themselves on the Moreno scale and how strongly 'British' they consider themselves to be.

If we examine the association between these two new scales measuring strength of Scottishness and strength of Britishness (Table 8.5), we find that a strong identity on one dimension is indeed somewhat associated with a weak identity on the other. Statistically, this is clear as the chi-square is significant at the 0.001 level. The association, however, is anything but perfect.[18] Nearly a quarter of those on point 6 of the Scottish scale choose points 6 and 7 on the British scale. Even among those who say they are very strongly Scottish three in ten do likewise. Clearly, the tendency to claim a British identity at, or stronger than, point 4, the mid-point of the scale, is apparent at all levels of Scottishness.

What is striking when we examine Table 8.5 further is that the largest proportion, 40 per cent, place themselves on points 5, 6 or 7 on both of the scales; they are 'dualists' in national identity terms, both strongly Scottish *and* strongly British. There is also a group which we might call 'nationalists'. These are the people who say they are 'strongly Scottish' (points 5, 6 or 7) but 'weakly British' (points 1, 2 or 3). They account for 28 per cent of the sample. There is a third group who are 'strongly Scottish' but who also sit at the mid-point (4) of the

[18] Using a nominal measure of association, phi = 0.499.

Table 8.5 *Strength of British and Scottish identity measured separately*

| Strength of Scottish identity (% by row) | Strength of British identity | | | | | | | | |
	1 Not at all British	2	3	4	5	6	7 Very strongly British	Base	Mean (SD)
1–4	2	7	5	59	5	4	18	99	4.4 (1.50)
5	1	5	8	37	36	8	5	139	4.47 (1.11)
6	2	9	16	24	23	18	6	248	4.34 (1.46)
7	16	10	11	21	12	6	23	483	4.15 (2.10)
All	9	9	11	28	18	9	16	969	4.27 (1.78)

Note: The first four points on the Scottish scale have been collapsed into one because only 18 people chose points 1–3 at the 'Not at all Scottish' end of the scale. Chi-square is highly significant (p < 0.001).

British scale, and who number 22 per cent. In other words, identity politics in Scotland takes place almost entirely among those who think of themselves as 'strongly Scottish' (points 5, 6 or 7), but especially between the 'nationalists' and the 'dualists'.

So who are these people who say they are strongly Scottish *and* strongly British, and how do they differ from the 'nationalists' who are strongly Scottish and weakly British? First of all, they tend to be women rather than men (43 per cent to 37 per cent), whereas the reverse is true for 'nationalists' (24 per cent to 32 per cent). They also tend to be older (43 per cent of people 65 or over), whereas 'nationalists' are proportionately young (42 per cent of those aged between 18 and 24). There are some modest social class and education differences such that 'dualists' are more likely to have had higher education and to be in higher managerial and professional employment, but the differences are not of real significance.

What is far more important in terms of making a difference are their 'political' and 'constitutional' preferences: how they voted in the 2011 Scottish Parliament election and how they wish Scotland to be governed.[19] 'Dualists' are less likely to have voted SNP and far less

[19] A binary logistic regression model including sex, age, education, social class, 2011 vote and constitutional preference shows that the two 'constitutional-political' variables are by far the most powerful ones determining a respondent's position on this new identity dimension.

likely to be in favour of independence. Those 'strong Scots' who are also 'strongly British' are making a powerful 'political-constitutional' statement that they can be both Scottish *and* British. The 'national-ists', strongly Scottish and weakly British, are firmly expressing their preference for an independent Scotland and voted disproportionately for the SNP at the 2011 Scottish election. We found the same pattern emerging when we replicated the analysis using the 2012 survey data. When we create a new kind of identity measure by using strength of Scottishness and Britishness we find a strong association with consti-tutional politics.

Before we set our discussion of 'national' and 'state' identities in a wider perspective, we must emphasise an important sociological point. How people choose to describe themselves in identity terms is a mat-ter of sociology rather than strict rules of logic. Who you say you are depends on the context in which you find yourself. Here is an example from a previous survey where we asked people in the Scottish *Gàidhealtachd*, notably Gaelic speakers, about their Scottish and their Gaelic identities. There, a number who said they were 'Gaels not Scots' on one scale, then said they were 'Scottish not British' on the other. Logically, of course, we would have expected them to deny that they were Scottish at all on the latter scale, having denied it on the former. What we inferred they were doing, however, was firmly asserting their Gaelic identity by opting for the 'Gael not Scots' category, the strong-est one available, but when faced with the question of being Scottish or being British, they were quite prepared to assert their Scottishness *in that different context*. Thus, if you have a strong sense of yourself as a Gael, but also as a Scot, it is sociologically meaningful to be a 'Gael not Scots' on one scale, and 'Scottish not British' on the other. Simply put, context matters.

The crisis of Britishness: the British state revisited

We began this chapter in elegiac mode: has being British had its day? There seemed, a priori, a good case for thinking that Scots, and maybe the English too for that matter, give little credence nowadays to their Britishness. Politicians, coming to similar conclusions, have rushed to the barricades, and in so doing, helped to draw attention to the apparent crisis of the British Union of which British identity seems an inextricable part. In his thoughtful book *Union and Unionisms*, the historian Colin Kidd observes: 'The status of the Union remains unresolved, but no longer is it – as was widely asserted to be prior to 1953 – something of a non-question' (2008: 133). It was another historian, Linda Colley, who gave earlier voice to the crisis of the British in her book *Britons: Forging the Nation* (1992). The use of the 'forging' metaphor, never quite made explicit by Colley, has two meanings: that of fusing two objects together to make one (in this case, nations); and second, a 'forgery', fabricating, somewhat unnaturally, a 'Britain' out of older nations. Colley observes: 'Britishness was superimposed over an array of internal differences in response to contact with the Other, and above all in response to conflict with the Other' (1992: 6). She points out that Britain

was an invention forged above all by war. Time and time again, war with France brought Britons, whether they hailed from Wales or Scotland or England, into confrontation with an obvious hostile Other and encouraged them to define themselves collectively against it. They defined themselves as Protestants struggling for survival against the world's foremost Catholic power. They defined themselves against the French as they imagined them to be, superstitious, militarist, decadent and unfree. (Colley, 1992: 6)

Colley's thesis has been influential, and was made so by the constitutional possibility of Britain having had its day. The end of Empire, the rise of political nationalism in Scotland and in Wales, the transformation into a 'multicultural' Britain especially in England in the post-war period, all raised the possibility of the 'end' of Britain, or at least its radical reinvention. Books with elegiac titles such as *After Britain* (Nairn, 2000) and *The Day Britain Died* (Marr, 2000) emerged at the turn of the new century. The erosion of the welfare state, the third pillar of Britishness after 'war' and 'religion', under the pressures from Thatcherism and the New Right, seemed to turn the screw. Given the absence of war, the decline of religion and the cutting back of the welfare state, which still continues, what was left to be 'British' about?

However, all was not quite what it seemed. As Kidd astutely observes, 'unionists' went to great lengths to assert the theoretical independence of Scotland, while 'nationalists' went to similar lengths to argue for equal treatment under that Union. The nub of Kidd's argument was that in Scotland at any rate, much of unionism is tinged with

nationalism, and much of nationalism with unionism. Unionism and
nationalism are intertwined. If we assume the contrary, that unionism
and nationalism are polar opposites, we are guilty, in Pierre Bourdieu's
words, of 'complicitous silence'.[20]

This, then, is the context in which much current debate about
'national identity' takes place. It reminds us that national identity is
malleable and varies as is appropriate according to context. It is not
a fixed measure to be presented on demand; which is, however, not to
say that people adopt *anything* which is to hand. It is a way of making
sense of the social and political world as they see it. Thus, the political
classes, both those seeking to maintain Union and those seeking independence, strain to separate 'national' from 'state' identity in the hope
that converting one to the other will pay political dividends for them.
The problem they face is, as we have tried to show in this chapter, that
this is not actually how people see it. Exhortations to 'just be Scottish',
or 'just be British' fall on deaf ears. Nor, even if this were not so, does
either identity translate easily into being in one constitutional camp or
the other. Simple binary divides between 'Scottish' and 'British' on the
one hand, and 'for' and 'against' Scottish independence on the other,
do not reflect the nuanced realities in which people lead their lives.
These are categories emanating from, and structured around, political
and cultural debates, which often do not correspond to the complexities and subtleties of people's lives. As the historian Dauvit Broun has
pointed out, assuming that Scottish independence and 'Britishness' are
incompatible – at any rate in fourteenth-century Scotland, which he
was talking about possibly tongue in cheek – is misguided.[21] In the
twenty-first century, it has become tediously necessary to repeatedly
point out, because many people seem so resistant to the idea, that one
cannot read off constitutional aspirations from how people construe
identities. It is far more complicated than that, and will continue so
to be. Even if Scotland becomes an independent state in the next few
years, it is unlikely that its citizens will wrap up their British identity

---

[20] The full quotation reads: 'The most successful ideological effects are those
which have no need of words, and ask no more than complicitous silence'
(quoted in Swartz, 2013: 101).

[21] Broun made the point in an article in *The Herald* newspaper, 25 February 2014
(www.heraldscotland.com/comment/columnists/agenda-historical-records-
show-earliest-scots-felt-they-were-real-britons.23500381; accessed 28
February 2014).

and send it, neatly packaged, to rUK[22] in the belief they have forfeited their claim to it. Political identities are too complex, too ragged, to be dealt with in this way.

Neither is it simply a 'political' matter. In sociological terms, our conclusions are only as good as our research instruments. We understand better than we once did that it is not a matter of choosing one identity over another, of making up your political mind once and for all. We have come to recognise the complexity and interrelationships of diverse territorial identities. It is the nature of the modern world that 'layering' and 'sharing' of powers and responsibilities are the norm, without implying that any simple 'nesting' of these is either desirable or even possible. Finally, just as 'the Union' of the twenty-first century is certainly not that of the early eighteenth, so 'Britain' is not the country it once was. Or is it? In a more recent book, *Acts of Union and Disunion* (2014), Linda Colley observes:

There was once a major panic about the health and future of Englishness. Some commentators argued it was Continental Europeans who were the problem, especially the French. Others blamed Scotland. Scots, they claimed, absorbed a disproportionate amount of Britain's tax revenue, while taking too many of its influential jobs, including that of prime minister. 'Into our places, states and beds they creep', wrote a particularly angry English poet of Scottish pretensions, 'They've sense to get, what we want sense to keep.' (Colley, 2014: 55)

No, not 2014, but 1760. *Plus ça change?*

---

[22] rUK has come to stand for the 'rest of the UK', comprising England, Wales and Northern Ireland, should Scotland leave the Union.

9 | *Whither national identity?*

In his classic 1961 paper entitled 'The oversocialized conception of man in modern sociology', Dennis Wrong began with the account of the writer Gertrude Stein on her deathbed. She is reported to have muttered: 'What, then, is the answer?' 'Pausing, she raised her head, murmured, "But what is the question?", and died.' (1961: 183). Wrong uses the story to make the point that answers are meaningless when separated from the questions:

> If we forget the questions, even while remembering the answers, our know-
> ledge of them will subtly deteriorate, becoming rigid, formal, and catechistic
> as the sense of indeterminacy, of rival possibilities, implied by the very put-
> ting of the question is lost. (Wrong, 1961: 183)

We shall conclude this final chapter with some speculative remarks about the future of research on national identity. The bulk of the chapter will pull together and expand on previous discussion of two key questions which, together, frame our approach to the study of national identity. First, why is national identity sociologically interesting? We argue that the concept helps us to address some classical questions about the nature of social order. The second question is: why have issues of national identity come to the fore in modern times in the British state? What does this tell us about the relationship between citizenship and 'nation-ness', notably in England and Scotland, and about the possible future of territorial relations in these islands? Here, then, are two fundamental questions pertaining to the social and polit- ical order which studying national identity can help us address.

Recovering the social

The first of these questions takes us back the classic problem of social order: how is society possible, or, in Dennis Wrong's words: 'How are

men capable of uniting to form enduring societies in the first place?'
Leaving aside his use of the male form, a feature of its time and not
ours, this remains arguably the central question in sociology. Wrong's
view is that answers to the question tend to focus too much on the
way social norms are internalised, which results in behaviour con-
forming to the social expectations of others and so creates a stable
society. What this kind of sociology produces is an oversocialised and
deterministic conception of human behaviour, rather than a properly
'social' conception.

   Inevitably, Wrong's classic paper is a creature of its time. He would
have considered it unthinkable that sociologists would not give the
'social' priority over 'the individual', whereas most sociologists today
accept that individual actors have a considerable amount of autonomy
and choice in their lives. Here is an example of the altered zeitgeist,
from Zygmunt Bauman:

Tell me what you buy and in which shops you buy it, and I'll tell you who
you are. It seems that with the help of carefully selected purchases I can
make of myself anything I may wish, anything I believe it is worth becoming.
Just as dealing with my personal problems is my duty and my responsibility,
so the shaping of my personal identity, my self-assertion, making myself into
a concrete someone, is my task and my task alone. (1996: 18–36)

Bauman may well have exaggerated for ironic effect, making the point
that this is how people see themselves, rather than what, in sociological terms, he believes, namely that they really are consuming creatures. The point remains, however, that the pendulum has swung a
long way from Wrong's concern that sociologists were treating human
beings as totally socialised. Sociology has moved from an overriding
focus on issues of 'structure' to concentrating more on 'action', to a
world in which doing 'identity politics' seems to be far more interesting to many scholars than how social, economic and political *structures* influence, even determine, that action. At its most extreme, there
is the post-modern, arguably non-sociological view that social actors
are entirely free agents so that social identities are highly liquid; that
we can be who we want to be by donning the appropriate garb. In the
last forty years, we have seen the rise and rise of 'individualism' politically, economically and culturally, and Wrong's question – how is
society possible – no longer seems central to many in the sociological

trade. On the other hand, perhaps, but only perhaps, the long reign of 'the individual' may be coming to an end, and Wrong's question becomes once again more salient and pressing.

Whatever the answer we may give, Wrong was correct in insisting that we start by asking the right questions. If we take the Hobbesian question of social control and social order as paramount, we might simply equate national identity with citizenship, which is laid down by society, or more precisely, its political agent, the state, and feel that there is little more to be said. That is effectively to rule out a more sociological argument by definition. On the other hand, if we assume that individuals are able to construct social identities for themselves, the central question of sociology once again raises its head. How does it come to pass that they do so in sufficiently similar and interlocking ways to make a stable society possible? The antitheses of constraint and choice, of structure and action, are as old as sociology itself but they still offer polarities within which to make sense of social identities, including national identities.

Scholars such as Benedict Anderson (1996) observed that discussing national identity is to examine the 'taken for granted', and this is implicit in some of what we shall say later in the chapter. In Chapter 6, we cited this comment by Mercer: 'Identity only becomes an issue when it is in crisis, when something assumed to be fixed, coherent and stable is displaced by the experience of doubt and uncertainty' (Mercer, 1990: 43). This helps us to problematise national identity in a novel way. What should be clear by now is that our approach to the study of national identity is profoundly sociological, premised on the centrality of 'society'. Reinstating the study of 'society', and how society is possible, as a central sociological issue seems to us long overdue (McCrone, 2010). Lest we are accused of adopting an 'oversocialised' conception which diminishes the active role of human beings, something which is far from our intentions, let us remind ourselves of Durkheim's stricture:

The characteristic attributes of human nature ... come to us through society. But on the other hand, society exists and lives only in and through individuals. Extinguish the idea of society in individual minds, let the beliefs, traditions, and aspirations of the collectivity cease to be felt and shared by the people involved, and society will die ... society has reality only to the extent that it has a place in human consciousness, and we make this place for it. (Durkheim, 2008 [1912]: 257)

The supreme theorist of the 'social', Durkheim had a proper understanding of the active role of the individual, and of social action. Reinstating 'society' seems essential, especially so in publications and discussions about nations, nationalism and, above all, national identity. Much of this general oeuvre has inadequately distinguished between the 'nation' and the 'state', the cultural and the political, and done so at the expense of the social. A society is a unit within whose boundaries social interaction is relatively dense and stable, and while some interactions take place across the boundaries, those taking place within them are the most significant and consistent. This is not to imply that associational life is internally homogeneous. It is, rather, in Michael Mann's words (1986), a 'loose confederation' containing 'overlapping networks of social interaction'.

Civil society refers to these relatively dense networks of organisations and institutions resulting from, and in turn, framing the day-to-day interactions of people. As Ernest Gellner (1994: 7–8) puts it, civil society is the social space located between the tyranny of kin and the tyranny of kings; between the intimacy of family life and the impersonal power of the state. It is intimately related to the 'state', the political level, but is not coterminous with it. Neither does civil society equate with the 'nation', for a sense of 'nation-ness' is sustained by institutional autonomy, rather than the other way round (McCrone, 2010: 184). The assertion of 'nationalism' derives not from elemental emotions based on historic memories, but from the day-to-day contemporary social associations of people. It arises from patterns of sociability structured by organisational life, what we might call 'civility'. A nation is an 'imagined community' *because* of its associational and institutional distinctiveness, and, as a result, it follows Durkheim's definition that: 'a society is not constituted by the mass who comprise it ... but above all by the idea it has of itself' (Fournier, 2013: 625–6). This sense of being a 'nation' is sustained by its sociality, and together, nation and society may well encourage a quest for some kind of state-like characteristics even if these fall short of what conventionally is called 'independence'.[1] There is considerable and repeated interaction between 'society', 'nation' and 'state' – the social, cultural and political levels, but they are distinct levels; we cannot subsume one into the others.

---

[1] Scotland at the time of writing exemplifies such a state-like situation, with its historically distinctive legal and educational system, together with many further powers delegated as a result of devolution.

Getting at national identity

We have taken the view throughout this book that national identity is an important and separate subject for study, but that many
scholars have wrongly treated it as the puny child of muscular
parents – nations and nationalism. It should by now be very clear
that we strongly dissent from the view that '"national identity"[2] [is]
a conceptual chimera not worthy of serous analytical pursuit. It is
a concept that is theoretically vapid while also lacking clear empirical referents' (Malesevic, 2011: 272–3). Why, if social actors use
'national', or any other form of social identity, to make sense of the
'objective material order', should any sociologist believe that it is not
worthy of comment and explanation? We should heed the classic dic-
tum of W. I. Thomas (2009 [1928]) that 'if people define situations as
real, they are real in their consequences'. The sociologist may think
or even know that the person doing the defining is mistaken, but
insofar as the error subsequently leads them to act in a particular
way, their definition of the situation has to be accepted. It would
have been of very little use, for example, for Leon Festinger and his
colleagues in their classic study *When Prophecy Fails* (1956) to point
out to the social actors who believed that the end of the world was
imminent that the 'objective material order' meant that the world
was not going to end after all. It was not their place as sociological
observers so to do, almost certainly would not have had the slightest
effect, and just might have deprived us of one of the classic studies
in social science.

Recall the comment made by the political philosopher David Miller
that national identity was to all intents and purposes unknowable: 'the
attitudes and beliefs that constitute nationality[3] [sic] are very often
hidden away in the deeper recesses of the mind, brought to full consciousness only by some dramatic event' (1995: 18). We have sympathy with the view that 'national identity' is implicit unless and until
mobilised by some salient event or stimulus not necessarily particularly dramatic, although whether it is located in the 'deeper recesses

[2] Note the quotation marks in the original text indicating that this author sees
the concept of national identity as *soi-disant*.
[3] It is worth repeating that in our view 'nationality' is too frequently taken in the
outside world as a synonym for citizenship for it to mean 'national identity'.

of the mind'[4] is to be questioned. In a later paper published in 2000, Miller appears to have moved away from that position and is prepared to use the Moreno categories of national identity. Writing of Belgium, the UK, Canada, Spain and Switzerland, he comments: 'two or more territorially-based communities exist within the framework of a single nation, so that members of each community typically have a split identity' (Miller, 2000: 129). More specifically, in Catalonia, 'we find that "Equally Spanish and Catalan" is the most popular self-description, followed by "More Catalan than Spanish", with smaller numbers claiming exclusively Catalan and exclusively Spanish identities' (Miller, 2000: 129–30).

Others hold to a more robustly sceptical view. In Chapter 1, we commented on the work of Brubaker and Cooper, who took issue with the concept of 'identity' in sociology, arguing that it explained everything and yet nothing, and hence was of limited analytical value. For Brubaker, identity is not a 'thing' which people have or do not have, or to which they aspire, and it cannot make them 'do' anything. We agree with Richard Jenkins' careful critique of this point of view, just as we too would reject the ontological status of 'identity', social or national. We too believe that our task is to carry out 'systematic inquiry into the observable realities of the human world' (Jenkins, 2008: 36). In any case, discarding the notion of national identity (with or without the quotation marks) is no solution:

It cannot really be done, if only because the genie is already out of the bottle. 'Identity' is not only an item in sociology's established conceptual toolbox; it also features in a host of public discourses, from politics to marketing to self-help. If we want to talk to the world outside academia, denying ourselves one of its words of power is not a good communications policy. (Jenkins, 2008: 14)

We find it helpful to use the metaphor of identity as the 'hinge' between social structure and social action. By this we mean that national identity is frequently the mechanism for connecting structure and action, and activating the latter. Only when something makes sense to those

---

[4] It is not clear exactly what Miller means by this phrase. It seems to go well beyond a sociological position such as our own or that of Benedict Anderson and to be closer to a psychoanalytical approach.

involved, and they then act on that basis (whether rightly or wrongly) do we get at sociological meaning. This is about more than social or national identity. For example, the classic trilogy in understanding social class is 'structure–consciousness–action', a title we used for an article with a colleague in the late 1970s on the British middle class (Bechhofer et al., 1978). The trilogy was not our invention, but common currency in studies of social stratification at the time. It made the point that 'structure' in and of itself does not lead directly to 'action', but has to be mediated by sets of meanings and understandings ('consciousness'). The failure to generate such common understandings among the French peasantry in the mid nineteenth century led Karl Marx in the *18th Brumaire of Louis Bonaparte* (1937 [1869]: ch. 7) to make his famous comment that peasants were akin to 'potatoes in a sack of potatoes', lacking the common awareness ('class consciousness') which could potentially lead to social and political action. Hence, it seems fitting to use 'national identity' (and other forms of social identity) in much the same way. Such a view is not confined to sociologists, but we adopt the famous saying of C. Wright Mills that 'every cobbler thinks leather is the only thing, and for better or worse, I am a sociologist' (1959: 19). Mills took the view that the 'sociological imagination' was not simply the preserve of sociologists, but 'the major common denominator of our cultural life and its signal feature' (1959: 14). This 'imagination' requires that we understand human beings as essentially social creatures, regardless of the prevailing individualistic wisdom.

The view that humans are thinking, sentient and, above all, social beings transcends narrow disciplinary boundaries. Here are three non-sociologists employing the sociological imagination. In Chapter 1, we quoted the late Neil MacCormick's incisive observation:

The truth about human individuals is that they are social products, not independent atoms capable of constituting society through a voluntary coming together. We are as much constituted by our society as it is by us. (MacCormick, 1999: 163)

Second, the social philosopher Alasdair MacIntyre observed that:

We all approach our own circumstances as bearers of a particular social identity. I am someone's son or daughter, someone's cousin or uncle; I am

a citizen of this or that city, a member of this or that guild or profession;
I belong to this clan, that tribe, this nation. Hence what is good for me has to
be the good for one who inhabits these roles. As such, I inherit from the past
of my family, my city, my tribe, my nation, a variety of debts, inheritances,
rightful expectations and obligations. These constitute the given of my life,
my moral starting point. (MacIntyre, 2007: 220)

And, third, the political scientist Michael Sandel reinforces the point:

If we understand ourselves as free and independent selves, unbound by
moral ties we haven't chosen, we can't make sense of a range of moral and
political obligations that we commonly recognize, even prize. These include
obligations of solidarity and loyalty, historic memory and religious faith –
moral claims that arise from the communities and traditions that shape our
identity. Unless we think of ourselves as encumbered selves, open to moral
claims we have not willed, it is difficult to make sense of these aspects of our
moral and political experience. (Sandel, 2009: 220)

The descriptor 'encumbered selves' makes our point. To paraphrase
Marx: we make ourselves, but not under conditions of our own making.[5] We see national identity as just one form of social identity among
many. It is pointless to try and prioritise one identity permanently
over another, as if there is a fixed hierarchy of such identities, as if
social class or 'race' or gender are always subordinated to national
identity, or vice versa. It all depends on how people choose to 'identify' themselves at key social moments, and for what purposes: actors
play the 'identity card' appropriate to the circumstances. Furthermore,
we think it necessary to make a distinction between the *content* of
national identity – what it is deemed to contain; its *boundaries* – who
is deemed inside and outside the collective 'we'; and its *salience* – how
people rate its importance to them in specific contexts.

When we reported in an earlier chapter on how respondents 'stacked
up' their social identities, the point was not to construct a hierarchy
of identities, but to get a sense of the salience of such identities in the
abstract. Thus, 'being Scottish' figured highly in the list for Scots, and

---

[5] The original quote is also from *The 18th Brumaire*: 'Men make their own
history, but they do not make it just as they please; they do not make it
under circumstances chosen by themselves, but under circumstances directly
encountered, given and transmitted from the past.'

even for the English 'national' identity, being English or British, was also salient. These are not the expression of some false consciousness.

If people choose (or not) to highlight their territorial identity, be it state or nation, that has to be accepted and taken at face value. It is not the outcome of some mindless exercise, but the outcome of a conscious process as to what is meaningful and appropriate in the circumstances. Neither is it purely 'emotional' and dissociated from material interests. Our work on 'claims' – who has or has not the right to be accepted on the basis of certain criteria – shows how the outcomes may connect with political concerns. If you are not 'one of us', then you do not have the right to benefit from our resources. The rise of right-wing parties throughout western Europe, and notably in Scandinavian countries seems linked to judgements as to who is or is not included as 'one of us' (see Mudde, 2007; Hainsworth, 2008; Keating and McCrone, 2013). We may deplore the mobilisation of 'identity' politics of this sort, but we cannot deny its salience and power. Nor is it simply confined to the politics of the racist Right. Such claims, and judgements about claims, are deeply embedded in mainstream political cultures. Although they are implicit most of the time, it takes little for them to be activated. Neither are such forms of 'identity politics' antithetical to 'material' politics, for they have the capacity to express and amplify such interests. To separate out 'identity politics' from 'real politics' (a term which usually means the analyst knows best) is a meaningless exercise. They are part of the same social phenomenon.

Politics and national identity

The second, and related, question we wish to discuss in this concluding chapter concerns the relationship between national identity and political-constitutional change in these islands, notably in Scotland and England. Indeed, after almost two decades of research, this has become, willy-nilly, the overarching context within which we have recently done much of our work. It is something of a double-edged sword. On the one hand, it is a recognisable context, while on the other, the elision of the political and the social, which cannot easily be separated, if at all, somewhat obfuscates our sociological interest.

These issues now largely dominate the domestic politics of the United Kingdom. The late Labour leader John Smith claimed that devolving power to a Scottish Parliament would come to represent the 'settled will' of the Scottish people. That will is no longer settled, if indeed it ever was. Indeed, the two constitutional referendums, on devolution in 1997 and on independence in 2014, are bookends for our research. Our work is on England and Scotland, and this focus can be justified intellectually because the Union of England and Scotland in 1707 was the basis of the UK state, the future of which is now in question, whereas neither Wales nor Northern Ireland have shown any substantial enthusiasm for secession rather than devolution.[6] National identity does not only encompass Scotland, England, Wales and Northern Ireland, because the term 'national identity' is used broadly to include what should strictly be called state identity – British – and this dimension is central to our work. Despite assumptions about the heightened salience of 'being British' (in 2012, both the Queen's Jubilee and the London Olympics took place), the effects, if they existed at all, appear to have been short-lived.[7] Over the *longue durée*, however, we do seem to be seeing a steady erosion of British claims, even among the English, though it would be wrong to conclude anything about either the rate of change or its eventual end state.

Our work does not have a prime 'political' focus, but politics should not be ignored or underplayed. Is national identity a good predictor of 'political' and constitutional developments? It is true that national identity has a *consistent* effect, going back at least twenty years, both on attitudes to Scottish self-government and on the propensity to vote for the Nationalist party but it is nothing like as clear-cut as one might imagine. Not everyone who thinks of themselves as 'Scottish not British' supports independence or the SNP, nor are 'Brits' (especially those who say they are more British than Scottish) opposed to greater Scottish self-government. Nonetheless, as we showed in the previous chapter, considering oneself 'strongly Scottish *and* strongly British' (as well as 'strongly Scottish *and* weakly British') is associated with how one voted in the 2011 Scottish election, and views on independence. Nor is the relationship between national identity and political-constitutional views only of historical interest. A run of key political events and elections are likely to determine the long-term shape of the UK: the 2014

---

[6] This is fortunate because we neither had the resources to study Wales, nor the expertise in Irish politics to study Northern Ireland.

[7] For example, in Scotland, the proportions saying they were 'mainly British' in 2012 and 2013 were virtually identical (11 per cent and 10 per cent respectively) (Curtice, 2014: table 3.1, p. 46).

Scottish referendum on independence; the 2015 UK election; and the 2016 Scottish Parliament election; and if the Conservative Party is re-elected, a promise of an in–out referendum on UK membership of the European Union. Those will need to be followed, documented and analysed. As much as anything, the sequence of events will matter, the one influencing the other in ways we cannot predict.

At the time of writing, before the Scottish independence referendum in September 2014,[8] it is impossible to know what the impact on 'national identities' will be in the longer term, and it will be many months, or even some years, before we can fully assess the impact. Should Scotland vote for independence, there will be complex and hotly disputed issues to be negotiated before full independence is declared, something which will take considerably longer than the eighteen months the Scottish Government envisage. The process will obviously be punctuated by a British general election in 2015, which would delay things further or indeed, especially if there is a change of government, lead to renegotiations on some issues. If Scotland votes 'no', the debates over the status quo versus further devolution will be extensive and complicated, and might well be even more prone to change after the British election.

A 'yes' vote would patently require new forms of 'citizenship' north and south of the border. We pointed out in the previous chapter that both an independent Scotland and the remaining UK (rUK) indicated that they would take a latitudinarian view of citizenship; thus, people born in Scotland but living elsewhere would have the right to apply for Scottish citizenship, just as those living in an independent Scotland would be eligible for British citizenship. But as we have repeatedly stressed, citizenship is *not* the same as national identity, which makes predictions about how national identity will unfold in these islands even more problematic. The advent of an independent Scotland would be a catalyst for change. Being 'British' is a clear-cut status if defined in citizenship terms, but not if, like being Scandinavian, it refers to geographical-historical-cultural matters.[9]

---

[8] We wish to remind readers that the entire manuscript with the exception of the Introduction is being written and submitted to Cambridge University Press well before the Scottish independence referendum, and our comments should be read accordingly.

[9] When asked whether, in the event of Scottish independence, they would still feel British 'due to geography, history and culture', 61 per cent of respondents surveyed agreed, and only 19 per cent disagreed (for details, see

If Scotland does not become independent, it is unlikely that the politics of national identity will cease. Indeed, they may take on greater urgency as territorial politics in the UK state are redefined, especially if there is a further referendum on UK membership of the European Union. If, for example, the majority of people in Scotland voted to stay in the EU, and the majority in the rest of the UK (but mainly England) voted to leave, then there would almost certainly be renewed debate about Scotland's membership of the UK. Certainly, the issues would become sharper and new battle lines drawn, especially if further powers are not devolved to the Scottish Parliament to keep it in the Union.

Is this obsession with national identity simply a parochial feature of these islands? Does anyone else care? It would seem so, for across western Europe at least, issues of national identity are prominent. For example, the knock-on effects of Scottish independence, and negotiations over EU membership of the new state,[10] would impact on Catalonia and Euskadi in Spain, something which is already reflected in nervousness in Brussels about issues of accession.[11] In a period of economic uncertainty and severe recession, questions about who we are and what we deserve become more salient. The crisis of the European Union is as much one of identity as it is of the economy. The revival of 'national' politics in a period of unprecedented and lengthy recession is a reflection of the fact that having 'made Europe', there has been a failure to make Europeans,[12] and hence there is merely a veneer of supranational identity on offer.

There is also a wider argument about the nature of 'nations' and 'nationalism' in the modern world. Craig Calhoun (2003) has argued that a false dichotomy has been set up between cosmopolitan liberalism on the one hand, and 'reactionary nationalism' on the other. The rhetorical opposition between the liberal cosmopolitan and the illiberal local remains influential (Ignatieff, 1999; Delanty and Kumar, 2006). Not only, in Calhoun's view, do the 'new cosmopolitans' have

whatscotlandthinks.org/questions/do-you-agree-or-disagree-that-if-scotland-becomes-independent-ill-feel-british#bar; accessed 11 July 2014).

[10] For discussion of the status of an independent Scotland in the EU, see www.bbc.co.uk/news/uk-scotland-scotland-politics-26173004 (accessed 10 July 2014).

[11] See www.bbc.co.uk/news/world-europe-25717161 (accessed 11 July 2014).

[12] The allusion is to the comment of the nineteenth-century Italian politician Massimo d'Azeglio: '*L'Italia è fatta. Restano da fare gli Italiani.*' ('We have made Italy, now we must make Italians.') For a review of 'European identity', see Miller and Day (2012).

no strong sense of social solidarity, they theorise about a world inhabited by autonomous, discrete and cultureless individuals, and as such
make common cause with economic liberals in dismissing the 'social'
as restrictive and potentially authoritarian:

At least in their extreme forms, cosmopolitanism and individualism partici-
pate in this pervasive tendency to deny the reality of the social. Their combination represents an attempt to get rid of 'society' as a feature of political
theory. It is part of the odd coincidence since the 1960s of left-wing and
right-wing attacks on the state. (Calhoun, 2003: 536)

Calhoun also makes the telling point that all identities and solidarities
are neither fixed nor simply fluid, but may become more or less fixed
under different circumstances and contexts. As Edensor observes: 'identity is always in process, is always being reconstituted in a process
of becoming and by virtue of location in social, material, temporal
and spatial contexts' (2002: 29). Identity is neither a 'thing' nor an
'essence' but a frame of reference for people. As we have previously
observed: 'It is both given to them, and made by them, in the course of
everyday interactions' (Bechhofer and McCrone, 2009b: 193).

The future of national identity research

We have dwelt thus far on one sense of the 'whither national identity' question. How it plays out on the ground will depend on how it
is mobilised in territorial politics. There is, however, another related
question: where should research on national identity now go? Bound
up together are matters of substance and of methods. Quite independently of current and future political developments, which have
their own momentum and rationale, the process of refining and
developing research instruments goes on. It may now seem obvious,
but only with hindsight, that as we have summarised above someone
can be strongly (or weakly) Scottish and strongly (or weakly) British.
It might seem surprising to some that there is continuing attachment to being British in Scotland. Why did we not think of exam-
ining that sooner? The short answer is that the so-called Moreno
scale had been widely used and served its purpose well, so well that
sociologists overlooked the fact that it led us to believe that being
British was weaker than it turned out to be. We make this point to

indicate that there is a close link between substantive findings and the methods used to obtain them, one informing the other. In this case, we were picking up a decline in the numbers saying 'British' because that was what the answers to the Moreno question indicated. Devising new ways of asking the national identity questions suggests that this is not entirely so in substantive terms. However, this most recent work relates to Scotland only and has not been carried out in England, where we know that national identity politics dance to a different tune.

Much has been made in the press about the rising tide of Englishness, and its possible political consequences, but the evidence thus far shows that if there is indeed a tide, it is remarkably slow-rising. Asking the equivalent two questions, about being strongly or weakly English and strongly or weakly British, may well reveal a different picture such that feeling strongly both is even more salient, but that is guesswork on our part. In any case, the simple demographic fact that people in England form 85 per cent of the UK population will lead to different national identity politics. Arguably, asserting 'national' over 'state' identity will have a different dynamic north and south of the border, especially as right-wing political parties such as the United Kingdom Independence Party (UKIP) have a much stronger presence in England than in Scotland.

The 'political-constitutional' aspects of national identity in these islands are an obvious point of interest, but there is much more to be investigated. We have been able to show that, contrary to expectations, national identity in Scotland as well as in England is 'cultural' rather than narrowly 'political', and that one's own sense of national identity is only loosely connected to how national and state symbols are identified (Chapter 6 and Bechhofer and McCrone, 2013). How national identity is imagined and mobilised with regard to 'race' and immigration is another understudied topic. For example, results from the 2011 Census for England[13] suggest that people who self-describe as belonging to ethnic minority groups are far more likely to describe themselves as 'exclusively British' than the white British population (by 38 per cent to 14 per cent). In Scotland, 26 per cent of ethnic

---

[13] 'Who Feels British? The relationship between ethnicity, religion and national identity in England', *Dynamics of Diversity: Evidence from the 2011 Census*: www.ethnicity.ac.uk/census/CoDE-National-Identity-Census-Briefing.pdf.

minorities said they were 'British only', while 34 per cent felt they had some Scottish identity, either on its own or in combination with another identity.[14] It is possible that such findings, taken on their own, are simply a function of how the question was asked, making our point about substance and method being bound up together.[15] Compared to the surveys on which we have reported throughout this book, the Census is an imperfect instrument for investigating national identity. The Census form is a self-completion document, and the 'responsible householder' is charged with ascribing a national identity to all household members. This raises serious question marks over its validity for this purpose, which take us back to the debates about the form in which questions about identity are asked (see Chapter 2).

Such findings, in truth, especially that 'white' people in England give far more precedence to being English than to being British, seem to contradict conventional academic wisdom on the issue. Krishan Kumar, whose book *The Making of English National Identity* (2003) is an intellectual tour de force, wrote subsequently:

What a mystery the English are, to themselves and others. Unlike their continental neighbours, the French, Germans and Italians, or even their nearest neighbours, the Welsh, Scottish and Irish, they have never established a strong sense of national identity. What is more, they have not shown much interest in inquiring, in any systematic way, into the character of themselves as a nation. (Kumar, 2011: xv)

The point of juxtaposing these apparently conflicting questions – do the English have a strong sense of their national identity or don't they – is to show that these are matters not easily resolved, and certainly open to further research.

What has been intriguing since Kumar wrote his book has been the emergence of scholarly interest in 'being English'. Susan Condor, who

[14] See www.scotlandscensus.gov.uk/news/census-2011-detailed-characteristics-ethnicity-identity-language-and-religion-scotland-; and for details, table DC2202SC – National identity by ethnic group.

[15] The Census question on national identity asked: 'How would you describe your national identity?' and people were encouraged to tick all that apply from English, Welsh, Scottish, Northern Irish, British and Other. Those ticking 'Other' were asked to write in their national identity. National identity is an exception to other questions in the Census because you are asked to describe rather than objectively state your status. The question also precedes questions on ethnic group, language, religion and passports held.

worked with us on the Leverhulme programme, has reinforced the point about the tendency to confuse national identity with nation and nationalism (Condor and Abell, 2006: 52). She has argued that the reticence of the English to talk about national identity does not mean they do not understand what the concept means (see Chapter 6 above for a fuller discussion of her perspective). What seems to us especially commendable is that Condor has carried out systematic *empirical* work on national identity in England. Her work has been influential, and duly acknowledged, in Michael Kenny's *The Politics of English Nationhood* (2014). Kenny's thesis is that:

there is a considerable body of evidence to support the conclusion that an avowed sense of English national identity has become more salient and meaningful for many people, and that this has developed at a greater distance from an established sense of allegiance to Britain. This emerging pattern of national identity may well turn out to constitute one of the most important phases in the history of the national consciousness of the English since the eighteenth century. (Kenny, 2014: 20–1)

Why that should be happening is discussed by Kenny, who does not attribute it directly to the asymmetrical system of devolution (for Scotland, Wales and Northern Ireland) introduced in 1997, but to factors such as the emergence of anti-system populism, and a reaction to multiculturalism (Kenny, 2014: 2). His study describes Englishness as akin to an '"empty signifier", which has been painted in various cultural and political colours and corralled in the service of a surprisingly wide range of arguments and ideas'[16] (Kenny, 2014: 6).

One such set of ideas, relating to resentment and linked to material and symbolic grievances, has been explored by Steve Fenton and Robin Mann, who observe:

In many cases this is characterised by grievances and a sense of unfairness, over the symbolic domain (e.g. flags, national holidays). This is evident through phrases such as 'why can't we be English?', 'we're not allowed to be English anymore', or 'if they're allowed to celebrate their culture why

---

[16] The concept 'empty signifier' (or 'floating signifier') refers to the lack of clear referents or substance, and entered social science from semiotics. It owed much to Claude Lévi-Strauss's discussion of the work of Marcel Mauss (Lévi-Strauss, 1987).

can't we?' and the common refrain of 'what about us?' (Fenton and Mann, 2013: 244)

Fenton (2012) characterises such respondents as 'resentful national-ists', echoing some of our own findings in Chapters 6 and 7. We are not claiming that the political rise of UKIP can be explained simply in terms of such resentments and that these lie behind expressions of being 'English'. There is, however, sufficient evidence to suggest, as Kenny does, that 'the notion that Englishness is, by definition, a "merely cultural", rather than political, phenomenon has led to an under-estimation of the different ways in which a renewed sense of Englishness is filtering into politics' (Kenny, 2014: 242).

The emergence of UKIP as a political force in England offers a good example of the ways in which the politics of identity and of social class come together. While they neither explore the significance of national identity, nor explicitly that of being English, Ford and Goodwin (2014) in their comprehensive study of UKIP reinforce the importance of the party's appeal. They point out that while the conventional wisdom is that UKIP draws mainly on disgruntled, middle-class and southern Conservatives,

UKIP's revolt is a working-class phenomenon. Its support is heavily con-centrated among older, blue-collar workers, with little education and few skills; groups who have been 'left behind' by the economic and social trans-formation of Britain in recent decades, and pushed to the margins as the main parties have converged in the centre ground. (Ford and Goodwin, 2014: 270)

One can see how the four key factors they identify in UKIP support – Euroscepticism, populism, opposition to migration, and economic and social pessimism – come to coalesce around a particular form of 'being English', and are amplified by rhetorical expressions of national identity.

Future-gazing is always a risky business, but it is possible that in ten years' time the important issues of 'national identity' in these islands will be construed as largely being 'about' England, rather than Scotland (or Wales and Northern Ireland, for that matter), whereas at present they are deemed to be about the so-called Celtic Fringe. We may well have moved on from the view that national identity is *not* about England, to one in which it is *all* about England. We would claim that one should never treat any national identity as an isolated and free-standing social form, because, as we argued in Chapter 7, 'notional others' are an integral part of the adoption of a national identity. In many respects, English nationalism is the 'elephant in the room' in these islands. We would not care to predict what cultural or political forms 'being English' might take, or what ramifications this would have throughout these islands. For example, there is an intriguing line of research developing around 'race' and ethnicity in England which argues that being 'white' becomes emblematically attached to 'working class' and stripped of association with 'middle class'. Lawler argues that social class and 'race' are doubly linked; that the working class 'are the bearers of a problematic whiteness, disavowed by liberal middle classes, and they are placed in a past, in a time before the rhetoric of multiculturalism, before such whiteness became a "problem" for white people themselves (its status has always been a problem for people who are not white)' (Lawler, 2012: 421).

This chimes with Fenton's evidence that scepticism about 'multicultural England/Britain' is more commonly found among the self-styled 'English', in contradistinction to those he terms 'liberal cosmopolitans' who 'may accept that they "are" English, but prefer the civic notion of being British' (Fenton, 2012: 466). Here we have a good example of how national identity categories are being used as cultural signifiers, especially where they imply racial/ethnic differences.

Then there are further issues of 'race' and immigration in these islands. For example, the in-migration of 'white' people from eastern Europe (notably from Poland) presents new opportunities for identity research. Are such people more (or less) likely to be thought of as 'one of us' than British-born people with non-white skin? Does the fact that they are not native-English speakers but 'look like' the natives make any difference to how the host society sees them? The calculus of 'race' vis-à-vis in-migration is as yet poorly understood. And there is the related set of issues as to whether 'new' migrants themselves feel able to claim new national identities at all easily, or tend to see their national identity in terms of their origins rather than their adopted land, always assuming that it is their intention to remain.

Another possible area for further work relates to comparing past and present. It is a commonplace to assume continuity from past to present, if only because nationalists (both political and cultural) see the past as something of a seamless web. In reality, there are of course major disjunctures, and one cannot assume that ancestors shared contemporary understandings of national identity, however much they are appealed to by later generations. Selective remembering (and forgetting – see Renan's celebrated 'remembering to forget' comment; Renan, 1990 [1882]: 11) tends to smooth out the contradictions and disjunctures between past and present. How it is that some people, events and processes are 'remembered' and appealed to while others are 'forgotten' remains an aspect of 'national identity' which is poorly understood.

We also need to know far more about what leads some social and cultural groups to make more 'national identity' claims than others, whether this is on the basis of social class, gender, ethnicity or regional variations. Neither should we think that only those with strong 'national identities' are worth studying. Those who deny that it has salience to them – 'cosmopolitans' – espouse different aspects of belonging and possibly none, and they are certainly not to be taken as the norm, the 'default', from which others are deemed to deviate. One might, for example, examine the hypothesis that elite groups do not attach much salience to national identity, regarding it as something which is for other – lower order – people (Calhoun, 2003; McCrone and Surridge, 1998).

National identity in a global world

Like Gertrude Stein, we might end up concluding that the questions are more interesting than the answers. Some people might argue that we are all global citizens now, and that localism, and nationalism, belong to another age. That would be to miss out some very important questions. If anything, the study of national identity has become more, not less, important. In the first place, it is contentious and maybe even wrong-headed to argue that the so-called nation-state has had its day. Global government seems very far off, and even 'regional' government along the lines of the European Union is in something of a crisis. By and large, it is to the state that citizens still look for solutions to problems. As Michael Mann pointed out:

We must beware the more enthusiastic of the globalists and transnationalists. With little sense of history, they exaggerate the former strength of nation-states; with little sense of global variety, they exaggerate their current decline; with little sense of their plurality, they downplay inter-national relations. (Mann, 1997: 494)

There are many instances, especially in western states, of 'stateless' or 'understated' nations that strive for greater political independence, such as Scotland, Catalonia, Quebec, Euskadi, Wales and Flanders (Rokkan and Urwin, 1982; Friend, 2012). The means of so doing – and such strivings may stop short of seeking formal independence – is by attachment to 'national' identity, to make the point that 'we' are culturally, socially and politically different from the state to which we are currently attached. That movement towards greater self-government has arguably been aided by the collapse of communism in the late 1980s and the end of the so-called Cold War, during which the opposing power 'blocs' had to be kept intact and their members corralled. With such constraints gone, pressure for greater autonomy, inside or outside the state, has grown on the back of a heightened sense of national identity, and is in turn helping to foster it. Analysts may find that threatening and dangerous, especially if they consider it 'thinking with the blood', but any review of 'neo-nationalism' shows that it is naive to simplify it to such an extent. Here, in these islands, we have our own version of such processes, but they are not peculiar to us. The relationship between 'national' and 'state' identities has not been resolved, and indeed, may never be. That is the condition of our times. The fact that people in their social and political struggles reach for national and for other forms of social identities shows that these are all meaningful in their lives, and that they act upon them, socially, politically and culturally. It behoves social scientists to make sense of that as best we can. As Richard Jenkins observes in his excellent review of social identity, 'Identification is a particularly seductive sociological topic because of the way in which it focuses the sociological imagination on the mundane dramas, dreams and perplexities of everyday human life' (2008: 16). To ask who 'we' are, and for what purposes, remains one of the key questions of our times.

Appendix: National identity publications

This appendix lists the main publications, mainly journal articles, from our work on national identity since 1998. It is arranged in reverse chronological order to allow the reader to follow the development of our ideas and techniques over the years.

'Who are we? Problematising national identity', in *The Sociological Review*, 46(4), 1998, pp. 629–52 (with R. Stewart and R. Kiely).

'Constructing national identity: arts and landed elites in Scotland', in *Sociology*, 33(3), 1999, pp. 515–34 (with R. Kiely and R. Stewart).

'Debatable land: national and local identity in a border town', in *Sociological Research Online*, vol. 5 (2), 2000 (www.socresonline.org.uk/5/2/kiely. html) (with R. Kiely and R. Stewart).

'The markers and rules of Scottish national identity', in *The Sociological Review*, 49(1), 2001, pp. 33–55 (with R. Kiely and R. Stewart).

'Keepers of the land: ideology and identities in the Scottish rural elite', in *Identities*, 8(3), 2001, pp. 381–409 (with R. Stewart and R. Kiely).

'Whither Britishness? English and Scottish people in Scotland', in *Nations and Nationalism*, 11(1), 2005, pp. 65–82 (with R. Kiely).

'Birth, blood and belonging: identity claims in post-devolution Scotland', in *The Sociological Review*, 53(1), 2005, pp. 150–71 (with R. Kiely).

'Reading between the lines: national identity and attitudes to the media in Scotland', in *Nations and Nationalism*, 12(3), 2006, pp. 473–92 (with R. Kiely).

'Being British: a crisis of identity?', in *The Political Quarterly*, 78(2), 2007, pp. 251–60.

'Talking the talk: national identity in England and Scotland', in *British Social Attitudes, 24th Report* (eds. A. Park et al.), 2008, pp. 81–104.

'National identity and social inclusion', in *Ethnic and Racial Studies*, 31(7), 2008, pp. 1245–66.

'Stating the obvious: ten truths about national identity', in *Scottish Affairs*, 67, spring 2009, pp. 7–22.

National Identity, Nationalism and Constitutional Change, F. Bechhofer and D. McCrone (eds.), 2009. Palgrave Macmillan. See especially

Ch. 1 'National identity, nationalism and constitutional change'; Ch. 4 'Being Scottish'; Ch. 9 'The politics of identity'.

'Claiming national identity', in *Ethnic and Racial Studies*, 33(6), 2010, pp. 921–48.

'Choosing national identity', in *Sociological Research Online*, 15(3), 2010.

'Imagining the nation: symbols of national culture in England and Scotland', in *Ethnicities*, 13(5), 2013, pp. 544–64.

'Changing claims in context: national identity revisited', in *Ethnic and Racial Studies*, 37(8), 2014, pp. 1350–70.

'The end of being British?', in *Scottish Affairs*, 23(3), 2014, pp. 309–31.

'What makes a Gael? Identity, language and ancestry in the Scottish Gàidhealtachd', in *Identities*, 21(2), 2014, pp. 113–33.

Bibliography

Alibhai-Brown, Yasmin 2000. *Who Do We Think We Are? Imagining the New Britain*. London: Penguin.

Anderson, Benedict 1996. *Imagined Communities: Reflections on the Origin and Spread of Nationalism*. London: Verso.

Aughey, Arthur 2012. 'Englishness as class: a re-examination', *Ethnicities*, 12(4): 394–408.

Aughey, Arthur and Berberich, Christine (eds.) 2011. *These Englands: A Conversation on National Identity*. Manchester University Press.

Bailey, F. G. 1996. *The Civility of Indifference: On Domesticating Ethnicity*. Ithaca, NY: Cornell University Press.

Banton, Michael 1983. *Racial and Ethnic Competition*. Cambridge University Press.

1999. *Ethnic and Racial Consciousness*, 2nd edn. London: Longman.

Barnett, Anthony 1997. *This Time: Our Constitutional Revolution*. London: Vintage.

2013. 'Beyond the "Global Kingdom": England after Scottish self-government', in Hassan and Mitchell, pp. 211–23.

Barth, Fredrik 1981. 'Ethnic group and boundaries', in *Process and Form in Social Life: Selected Essays of Fredrick Barth: Volume 1*. London: Routledge and Kegan Paul, pp. 198–227.

Bauman, Zigmunt 1996. 'From pilgrim to tourist: – or a short history of identity', in Hall and DuGay (eds.), pp. 18–36.

Bechhofer, Frank, Elliott, Brian and McCrone, David 1978. 'Structure, consciousness and action: a sociological profile of the British middle class', *British Journal of Sociology*, 29(4): 410–36.

Bechhofer, Frank and McCrone, David 2008. 'Talking the talk: national identity in England and Scotland', in Park et al. (eds.), pp. 81–104.

2009a. 'Stating the obvious: ten truths about national identity', *Scottish Affairs*, 67, spring: 7–22.

2009c. 'Being Scottish', in Bechhofer and McCrone (eds.), pp. 64–94.

2010. 'Choosing national identity', *Sociological Research Online*, 15(3). (www.socresonline.org.uk/15/3/3.html; accessed 3 March 2014).

2013. 'Imagining the nation: symbols of national culture in England and Scotland', *Ethnicities*, 13(5): 544–64.

2014a. 'Changing claims in context: national identity revisited', *Ethnic and Racial Studies*, 37(8): 1350–70.

2014b. 'The end of being British?', *Scottish Affairs*, 23(3): 309–31.

(eds.) 2009b. *National Identity, Nationalism and Constitutional Change*. London: Palgrave Macmillan.

Bechhofer, Frank and Paterson, Lindsay 2000. *Principles of Research Design in the Social Sciences*. London: Routledge.

Beiner, Robert (ed.) 1999. *Theorizing Nationalism*. Albany, NY: SUNY Press.

Berger, Stefan 2009. 'On the role of myths and history in the construction of national identity in modern Europe', *European History Quarterly*, 39(3): 490–502.

Billig, Michael 1995. *Banal Nationalism*. London: Sage Publications.

2009. 'Reflecting on a critical engagement with banal nationalism – reply to Skey', *The Sociological Review*, 57(2): 347–52.

Bogdanor, Vernon 2013. 'The one-state solution to England's role in a devolved UK', *The Guardian*, 8 April (www.theguardian.com/commentis free/2013/mar/25/one-state-solution-england-devolved-uk?INTCMP= SRCH&guni=Article:in%20body%20link; accessed 3 April 2014).

Bond, Ross and Paterson, Lindsay 2005. 'Coming down from the ivory tower? Academics' civic and economic engagement with the community', *Oxford Review of Education*, 31: 331–51.

Bourdieu, Pierre 1984. *Distinction: A Social Critique of the Judgement of Taste*. London: Routledge and Kegan Paul.

Brown, Gordon 2010. *'Being British: The Search for the Values that Bind the Nation'* (with Matthew d'Ancona). Edinburgh: Mainstream.

2014. *My Scotland, Our Britain: A Future Worth Sharing*. London: Simon and Schuster.

Brubaker, Rogers and Cooper, Frederick 2000. 'Beyond "Identity"', *Theory and Society*, 29(1): 1–47.

Bulmer, Elliot. 2013. 'The Scottish referendum is about popular sovereignty, not identity', *The Guardian*, 31 March (www.theguardian.com/commentisfree/2013/mar/31/scotland-referendum-sovereignty-identity-bogdanor; accessed 3 April 2014).

Calhoun, Craig 2003. '"Belonging" in the cosmopolitan imagery', *Ethnicities*, 3(4): 531–68.

Chesterton, G. K. 1915. *The Secret People* (www.cse.dmu.ac.uk/~mward/gkc/books/secret-people.html; accessed 13 March 2014).

Cohen, Anthony 1994. *Self Consciousness: An Alternative Anthropology of Identity*. London: Routledge.

2000. *Signifying Identities: Anthropological Perspectives on Boundaries and Contested Values*. London: Routledge.

Cohen, Robin 1994. *Frontiers of Identity: The British and Others*. London: Longman.

Colley, Linda 1992. *Britons: Forging the Nation 1707–1837*. London: Yale University Press.

2014. *Acts of Union and Disunion*. London: Profile Books.

Condor, Susan 2006. 'Temporality and collectivity: diversity, history and the rhetorical construction of national entitativity', *British Journal of Social Psychology*, 45: 657–82.

2010. 'Devolution and national identity: the rules of English (dis)engagement', *Nations and Nationalism*, 16(3): 525–43.

2011. 'Sense and sensibility', in Aughey and Berberich (eds.), pp. 29–55.

2012. 'Understanding English public reactions to the Scottish Parliament', *Nations and Nationalism*, 14(1): 83–98.

Condor, Susan and Abell, Jackie 2006. 'Vernacular constructions of "national identity" in post-devolution Scotland and England', in Wilson and Stapleton (eds.), pp. 51–76.

Condor, Susan and Fenton, Steve 2012. 'Thinking across domains: class, nation and racism in England and Britain', *Ethnicities*, 12(4): 385–93.

Condor, Susan and Gibson, Stephen 2007. '"Everybody's entitled to their opinion": ideological dilemmas of liberal individualism and active citizenship', *Journal of Community and Applied Social Psychology*, 17: 115–40.

Condor, Susan, Gibson, Stephen and Abell, Jackie 2006. 'English identity and ethnic diversity in the context of UK constitutional change', *Ethnicities*, 6: 123.

Copus, Colin 2011. 'Englishness and local government: reflecting a nation's past or merely an administrative convenience?', in Aughey and Berberich (eds.), pp. 193–213.

Crick, Bernard 1989. 'An Englishman considers his passport', in Evans (ed.), pp. 23–44.

2010. 'The four nations: interrelations', *Scottish Affairs*, no. 71, spring: 3–15.

Crouch, Colin 2013. 'Class politics and the social investment welfare state', in Keating and McCrone (eds.), pp. 156–68.

Curtice, John 2011. 'Is the English lion ready to roar?', in Aughey and Berberich (eds.), pp. 56–74.

2014. 'Independence referendum: a question of identity, economics or equality?', in *British Social Attitudes*, 31, NatCen Social Research, pp. 42–60 (www.natcen.ac.uk/our-research/research/british-social-attitudes/; accessed 11 July 2014).

Curtice, John and Heath, Anthony 2009. 'England awakes? Trends in national identity in England', in Bechhofer and McCrone (eds.), pp. 41–63.

Davies, Christie 1990. *Ethnic Humor Around the World: A Comparative Analysis*. Bloomington, IN: Indiana University Press.

Delanty, Gerard and Kumar, Krishan (eds.) 2006. *The Sage Handbook of Nations and Nationalism*. London: Sage Publications.

Durkheim, Emile 2008 [1912]. *Elementary Forms of Religious Life*. Oxford University Press.

Edensor, Tim 2002. *National Identity, Popular Culture and Everyday Life*. Oxford: Berg.

Eriksen, T. H. 1993. *Ethnicity and Nationalism: Anthropological Perspectives*. London: Pluto Press.

Evans, Neil (ed.) 1989. *National Identity in the British Isles*, Occasional Papers in Welsh Studies, no. 3. Centre for Welsh Studies, Coleg Harlech, Gwynedd.

Fenton, Steve 2012. 'Resentment, class and social sentiments about the nation: the ethnic majority in England', *Ethnicities*, 12(4): 465–83.

Fenton, Steve and Mann, Robin 2013. '"Our own people": ethnic majority orientations to nation and country', in T. Modood and J. Salt (eds.) *Global Migration, Ethnicity and Britishness*. London: Palgrave Macmillan.

Festinger, Leon, Riecken, Henry and Schachter, Stanley 1956. *When Prophecy Fails*. Minneapolis: University of Minnesota Press.

Ford, Robert and Goodwin, Matthew 2014. *Revolt of the Right: Explaining Support for the Radical Right in Britain*. London: Routledge.

Fournier, Marcel 2013. *Emile Durkheim*. London: Polity Press.

Fox, Jon E. and Miller-Idriss, Cynthia 2008a. 'Everyday nationhood', *Ethnicities*, 8(4): 536–62.
  2008b. 'The "here and now" of everyday nationhood', *Ethnicities*, 8(4): 573–6.

Fraser, George Macdonald 1971. *The Steel Bonnets: The Story of the Anglo-Scottish Border Reivers*. London: Harper Collins.

Friend, Julius 2012. *Stateless Nations: Western European Regional Nationalisms and the Old Nations*. London: Palgrave Macmillan.

Geertz, Clifford 1973. *The Interpretation of Cultures: Selected Essays*. New York: Basic Books.

Gellner, Ernest 1964. *Thought and Change*. London: Weidenfeld and Nicolson.
  1973. 'Scale and Nation', *Philosophy of the Social Sciences*, 3: 1–17.
  1994. *Conditions of Liberty: Civil Society and its Rivals*. London: Hamish Hamilton.

Glauser, Beat 1974. *The Scottish-English Linguistic Border: Lexical Aspects*. Bern: Franck.

Goffman, Ernest 1959. *The Presentation of Self in Everyday Life*. New York: Doubleday Anchor Books.

Goulbourne, Harry 1991. *Ethnicity and Nationalism in Post-Imperial Britain*. Cambridge University Press.

Greig, Andrew 2013. *Fair Helen*. London: Quercus Books.

Hainsworth, Paul 2008. *The Extreme Right in Western Europe*. London: Routledge.

Hall, John A. 1998. *The State of the Nation: Ernest Gellner and the Theory of Nationalism*. Cambridge University Press.

Hall, Stuart 1995. 'Negotiating Caribbean identities', *New Left Review*, 209 (January–February): 3–14.

Hall, Stuart and DuGay, Paul (eds.) 1996. *Questions of Cultural Identity*. London: Sage.

Hassan, Gerry and Mitchell, James (eds.) 2013. *After Independence*. Edinburgh: Luath Press.

Hazell, Robert (ed.) 2006. *The English Question*. Manchester University Press.

Hearn, Jonathan 2000. *Claiming Scotland: National Identity and Liberal Culture*. Edinburgh: Polygon.
   2009. 'Small fortunes: nationalism, capitalism and changing identities', in Bechhofer and McCrone (eds.), pp. 144–62.

Heffer, Simon 1999. *Nor Shall My Sword: The Reinvention of England*. London: Weidenfeld and Nicolson.

Hobsbawm, Eric 1977. 'Some reflections on "The Break-Up of Britain"', *New Left Review*, 105 (September–October): 3–23.

Hussain, Asifa and Miller, William 2006. *Multicultural Nationalism: Islamophobia, Anglophobia, and Devolution*. Oxford University Press.

Ignatieff, Michael 1999. 'Nationalism and the narcissism of minor differences', in Beiner (ed.), pp. 91–102.

Jeffery, Charlie and Mitchell, James (eds.) 2009. *The Scottish Parliament: The First Decade*. Edinburgh: Luath Press in association with Hansard Society.

Jenkins, Richard 2008. *Social Identity*, 3rd edn. London: Routledge.

Jones, Richard Wyn, Lodge, Guy, Henderson, Ailsa, and Wincott, Daniel 2012. *The Dog that Finally Barked: England as an Emerging Political Community*. London: IPPR.

Jowell, Roger, Jowell, R., Curtice, J., Park, A., Brook, L., Thomson, K., and Bryson, C. (eds.) 1998. *British and European Social Attitudes: The 15th Report: How Britain Differs*. Aldershot: Ashgate.

Judah, Tim 1997. *The Serbs: History, Myth and the Destruction of Yugoslavia*. London: Yale University Press.

Kapferer, Bruce 1988. *Legends of People, Myths of State: Violence, Intolerance and Political Culture in Sri Lanka and Australia*. Washington, DC: Smithsonian Institution Press.

Keating, Michael and McCrone, David (eds.) 2013. *The Crisis of Social Democracy in Europe*. Edinburgh University Press.

Kenny, Michael 2014. *The Politics of English Nationhood*. Oxford University Press.

Kidd, Colin 2008. *Union and Unionisms: Political Thought in Scotland, 1500–2000*. Cambridge University Press.

Kiely, Richard, McCrone, David, Bechhofer, Frank and Stewart, Robert 2000. 'Debatable land: national and local identity in a border town', *Sociological Research Online*, 5(2). Available at www.socresonline.org.uk/5/2/kiely.html.

Kiely, Richard, McCrone, David and Bechhofer, Frank 2006. 'Reading between the lines: national identity and attitudes to the media in Scotland', *Nations and Nationalism*, 12(3): 473–92.

Kohn, Hans 1944. *The Idea of Nationalism: A Study in its Origins and Background*. New York: Macmillan.

Kumar, Krishan 2003. *The Making of English National Identity*. Cambridge University Press.

2006. 'English and French national identity: comparisons and contrasts', *Nations and Nationalism*, 12(3): 413–32.

2011. Preface, in Aughey and Berberich (eds.), pp. xv–xvi.

Kymlicka, Will 2006. 'The evolving basis of European norms of minority rights: rights to culture, participation and autonomy', in McGarry and Keating (eds.), pp. 35–63.

Lawler, Steph 2012. 'White like them: whiteness and anachronistic space in representations of the English white working class', *Ethnicities*, 12(4): 409–26.

Lazarsfeld, Paul 1949. 'The American Soldier – an expository review', *The Public Opinion Quarterly*, 13(3): 377–404.

Lee, Simon 2007. *Boom and Bust: The Politics and Legacy of Gordon Brown*. Oxford: Oneworld.

2011. 'Gordon Brown and the negation of England', in Aughey and Berberich (eds.), pp. 155–73.

Lévi-Strauss, Claude 1987. *Introduction to the Work of Marcel Mauss*. London: Routledge.

MacCormick, Neil 1996. 'Liberalism, nationalism and the post-sovereign state', *Political Studies*, 44: 553–67.

1999. *Questioning Sovereignty*. Oxford University Press.

McCrone, David 2001. *Understanding Scotland: The Sociology of a Nation,* 2nd edn. London: Routledge.

2010. 'Recovering civil society: does sociology need it?', in P. Baert, S. Koniordos, G. Procacci and C. Ruzza (eds.) *Conflict, Citizenship and Civil Society,* pp. 183–99. London: Routledge.

McCrone, David and Bechhofer, Frank 2010. 'Claiming national identity', *Ethnic and Racial Studies,* 33(6): 921–48.

McCrone, David and Surridge, Paula 1998. 'National identity and national pride', in Jowell et al. (eds.), pp. 1–18.

McGarry, John and Keating, Michael (eds.) 2006. *European Integration and the Nationalities Question.* London: Routledge.

MacIntyre, Alasdair 2007. *After Virtue: A Study in Moral Theory.* London: Duckworth.

Mackenzie, W. J. M. 1978. *Political Identity.* Manchester University Press.

Malcolm, Noel 1994. *Bosnia: A Short History.* London: Macmillan.

Malesevic, Sinisa 2011. 'The chimera of national identity', *Nations and Nationalism,* 17(2): 272–90.

Malesevic, Sinisa and Haugaard, M. (eds.) 2007. *Ernest Gellner and Contemporary Social Theory.* Cambridge University Press.

Mandler, Peter 2006. 'What is "national identity"? Definitions and applications in modern British historiography', *Modern Intellectual History,* 3(2): 271–97.

Mann, Michael 1986. *The Sources of Social Power, Volume 1: A History of Power from the Beginning to A.D. 1760.* Cambridge University Press.

1997. 'Has globalization ended the rise and rise of the nation-state?', *Review of International Political Economy,* 4(3): 472–96.

Mann, Robin 2012. 'Uneasy being English: the significance of class for English national sentiment', *Ethnicities,* 12(4): 484–99.

Marr, Andrew 2000. *The Day Britain Died.* London: Profile Books.

Marti, David 2013. 'The 2012 Catalan Election: the first step towards independence?', *Regional and Federal Studies,* 23(4): 507–16.

Marx, Karl 1937 [1869]. *The Eighteenth Brumaire of Louis Napoleon.* Moscow: Progress Publishers.

Meek, James 2014. 'The leopard', in *London Review of Books,* 19 June.

Mercer, K. 1990. 'Welcome to the jungle', in Rutherford (ed.), pp. 43–71.

Meret, Susie and Siim, Birte 2013. 'Multiculturalism, right-wing populism and the crisis of social democracy', in Keating and McCrone (eds.), pp. 125–39.

Miller, David 1995. *On Nationality.* Oxford: Clarendon Press.

2000. *Citizenship and National Identity.* Cambridge: Polity Press.

Miller, Robert and Day, Graham (eds.) 2012. *The Evolution of European Identities: Biographical Approaches.* London: Palgrave Macmillan.

Mills, C. Wright 1959. *The Sociological Imagination*. London: Oxford University Press.

Modood, Tariq 2010. *Still Not Easy Being British: Struggles for a Multicultural Citizenship*. Stoke-on-Trent: Trentham Books.

Moore, Margaret 2006. 'Nationalism and Political Philosophy', in Delanty and Kumar (eds.), pp. 94–103.

Moreno, Luis 2006. 'Scotland, Catalonia, Europeanization and the "Moreno Question"', *Scottish Affairs*, 54: 1–21.

Mudde, Cass 2007. *Populist Radical Right Parties in Europe*. Cambridge University Press.

Nairn, Tom 1977. *The Break-Up of Britain: Crisis and Neo-Nationalism*. London: New Left Books.

1988. 'The Three Dreams of Scottish Nationalism', in Paterson (ed.), pp. 31–9.

2000. *After Britain: New Labour and the Return of Scotland*. London: Granta Publications.

O'Dowd, Liam and Wilson, Tom 1996. *Borders, Nations and States: Frontiers of Sovereignty in the New Europe*. Aldershot: Avebury.

O'Leary, Brendan (ed.) 1996. 'A Symposium on David Miller's *On Nationality*', *Nations and Nationalism*, 2(3): 407–51.

Ormston, Rachel and Curtice, John 2012. *The English Question: How is England Responding to Devolution?* London: NatCen Social Research. Available at: www.bsa-29.natcen.ac.uk/read-the-report/scottish-independence/does-scotland-want-independence.aspx (accessed 4 March 2014).

Park, A., Curtice, J., Thomson, K., Phillips, M., Johnson, M., and Clery, E. (eds.) 2008. *British Social Attitudes, 24th Report*. London: Sage.

Park, Robert, and Burgess, Ernest W. 1984 [1925]. *The City*. University of Chicago Press.

Paterson, Lindsay (ed.) 1988. *A Diverse Assembly: The Debate on a Scottish Parliament*. Edinburgh University Press.

2015. 'Utopian pragmatism: Scotland's choice', *Scottish Affairs*, 24(1): 22–46.

Petersoo, Pille 2007. 'What does "we" mean? National deixis in the media', *Journal of Language and Politics*, 6(3): 419–36.

Pocock, J. G. A. 1975. 'British History: a plea for a new subject', *The Journal of Modern History*, 47(4): 601–21.

Rapport, Nigel 2008. *Of Orderlies and Men: Hospital Porters Achieving Wellness at Work*. Durham, NC: Carolina Academic Press.

Reicher, Stephen and Hopkins, Nick 2001. *Self and Nation: Categorization, Contestation and Mobilization*. London: Sage Publications.

Reicher, Stephen, McCrone, David and Hopkins, Nick 2010. '"A strong, fair and inclusive national identity": a viewpoint on the Scottish Government's Outcome 13', *Equality and Human Rights Commission Research Report 62* (www.equalityhumanrights.com/uploaded_files/research/national_identity_viewpoint_research_report_62.pdf; accessed 3 March 2014).

Reicher, Steve, Cassidy, Clare, Wolpert, Ingrid, Hopkins, Nick and Levine, Mark 2006. 'Saving Bulgaria's Jews: an analysis of social identity and the mobilisation of social solidarity', *European Journal of Social Psychology*, 46: 49–72.

Renan, Ernest 1990 [1882]. 'What is a nation?', reprinted in H. K. Bhabha (ed.), *Nation and Narration*. London: Routledge, pp. 8–22.

Revue internationale de politique comparée, 2007. 'La concurrence des identités: débats à propos de l'utilisation de la Question Moreno', 14(4): 489–612.

Rokkan, Stein and Urwin, Derek 1982. *The Politics of Territorial Identity: Studies in European Regionalism*. London: Sage.

Rutherford, John (ed.) 1990. *Identity: Community, Culture, Difference*. London: Lawrence and Wishart.

Sallis, James 2008. *Salt River*. Harpenden: No Exit Press.

Sandel, Michael 2009. *Justice: What's the Right Thing To Do?* London: Penguin.

Scruton, Roger 1980. *The Meaning of Conservatism*. London: Macmillan.

Skey, Michael 2009. 'The national in everyday life: a critical engagement with Michael Billig's thesis of *Banal Nationalism*', *The Sociological Review*, 57(2): 331–46.

Smith, Anthony 1991. *National Identity*. Harmondsworth: Penguin.
    2008. 'The limits of everyday nationhood', *Ethnicities*, 8(3): 563–72.

Swartz, David L. 2013. *Symbolic Power and Intellectuals: The Political Sociology of Pierre Bourdieu*. London: University of Chicago Press.

Tamir, Yael 1993. *Liberal Nationalism*. Princeton University Press.

Taylor, Peter 1993. 'The meaning of the North: England's "foreign country" within?', *Political Geography*, 12(2): 136–55.

Thomas, William Isaac 2009 [1928]. *The Child in America: Behavior Problems and Programs*. New York: Knopf.

Watt, Dominic, Llamas, Carmen and Johnson, Daniel 2010. 'Levels of linguistic accommodation across a national border', *Journal of English Linguistics*, 38(3): 270–89.

Wetherell, Margie 2009. *Theorizing Identities and Social Action*. London: Palgrave Macmillan.

Wilson, John and Stapleton, Karyn (eds.) 2006. *Devolution and Identity*. Aldershot: Ashgate.

Wodak, Ruth, de Cillia, Rudolf, Reisigl, Martin and Liebhart, Karin 1999. *The Discursive Construction of National Identity*. Edinburgh University Press.

Worsley, Peter 1984. *The Three Worlds: Culture and World Development*. London: Weidenfeld and Nicolson.

Wright, Patrick 2005. 'Last Orders', in *The Guardian*, 9 April. Available at www.theguardian.com/books/2005/apr/09/britishidentity.society, accessed 13 March 2014.

Wrong, Dennis 1961. 'The oversocialized conception of man in modern sociology', *American Sociological Review*, 26(2): 183–93.

Yack, Bernard 2012. *Nationalism and the Moral Psychology of Community*. University of Chicago Press.

Index

Abell, Jackie, 23, 54, 67
accent
  Berwick-upon-Tweed, 85–6
  English nationality, race, place of
    birth and accent, 107–11
  as marker, 29, 100–1
  and non-white citizens, 105, 106
  questions, Social Attitudes
    Surveys, 38–9
  Scottish nationality, race, place of birth
    and accent, 106–7, 111–12
age, as context for national
  identities, 66
Alnwick
  Berwickers seen as Scottish by, 84–7
  denial of a Geordie identity, 86
  English identity in, 90
  regional identification and, 84
Amis, Martin, 150
Anderson, Benedict, 10–11, 190
Anglophobia, 153

Bailey, F. G., 161–2
Balkans conflict
  ethnicity in, 115, 143, 161
  Yugoslavia, conflict-prone borders, 71
Banton, Michael, 29, 142, 143, 145
Barnett, Anthony, 45, 168
Barth, Frederik, 143
Bauman, Zygmunt, 189
Berger, Stefan, 70
Berwick Rangers, 94
Berwickers
  Berwicker identity over national
    identity, 74–7, 78
  dialect and, 77
  English identity of, 73
  as English, Eyemouth
    perception, 87–8
  Scottish identity of, 80

as Scottish, Alnwickers'
  perception, 84–7
  term as an insult, Eyemouth, 89
  views of, neighbouring English
    villages, 81
  views of, neighbouring Scottish
    villages, 81–2
Berwick-upon-Tweed
  accent, 85–6
  birthplace, importance of, 7,
    83, 87–8
  as border town, 70
  contested identity, research
    into, 28
  devolution, concerns over, 92, 96
  distinctive national identities in,
    68, 73, 83
  English identity in, 72–3, 78, 81–2
  family/ancestry and, 76, 77, 79–80
  identity for organisations, 93–5
  incomers to, 79–80
  institutional/jurisdictional ambiguity,
    82–3, 91–3
  localism over national identity, 74–7,
    78, 82, 88–9, 90–1
  regional identification and, 75–6,
    84
  rivalry with Eyemouth, 89–90
  Scottish identity in, 71–4, 77–8
  war with Russia, 69–70
Billig, Michael, 7, 13, 16, 71, 96
birthplace
  Berwick-upon-Tweed and
    nationality, 7, 83, 87–8
  and claims to national identities,
    99–100, 101–2, 106
  English nationality, race, place of
    birth and accent, 107–11
  as identity marker, 27, 30, 98–100,
    101–2, 106

220

Printed in the United States
By Bookmasters